Springer Series in Advanced Microelectronics 28

The Springer Series in Advanced Microelectronics provides systematic information on all the topics relevant for the design, processing, and manufacturing of microelectronic devices. The books, each prepared by leading researchers or engineers in their fields, cover the basic and advanced aspects of topics such as wafer processing, materials, device design, device technologies, circuit design, VLSI implementation, and subsystem technology. The series forms a bridge between physics and engineering and the volumes will appeal to practicing engineers as well as research scientists.

For other titles published in this series, go to
www.springer.com/series/4076

Michael Fulde

Variation Aware Analog and Mixed-Signal Circuit Design in Emerging Multi-Gate CMOS Technologies

Springer

Michael Fulde
Infineon Technologies Austria AG
Siemensstr. 2
9500 Villach
Austria
Michael.Fulde@infineon.com

ISSN 1437-0387
ISBN 978-94-007-3083-0 e-ISBN 978-90-481-3280-5
DOI 10.1007/978-90-481-3280-5
Springer Dordrecht Heidelberg London New York

Cover design: eStudio Calamar S.L.

Printed on acid-free paper

Springer is part of Springer Science+Business Media (www.springer.com)

Preface

The commercial success of semiconductor industry is mainly driven by the continuous scaling of CMOS and the proceeding functional integration in system on chip applications. Reaching the nanometer scale severe scaling limitations enforce the introduction of novel materials, device architectures and device concepts. Multi-gate FETs employing high-k gate dielectrics are considered as promising solution overcoming the scaling limitations of conventional planar bulk CMOS. Especially analog, mixed-signal and RF device and circuit performance is affected by these revolutionary changes in technology. This work provides a technology oriented assessment of analog and mixed-signal circuits in emerging multi-gate CMOS technologies. On device level, the reduction of short channel effects is a major advantage of fully depleted multi-gate devices, resulting in beneficial output impedance, gain and matching behavior. Serious concerns related to high-k dielectrics are pronounced flicker noise and dynamic threshold voltage variations or hysteresis effects. The impact of flicker noise on circuits and noise reduction techniques are briefly discussed. A model for hysteresis effects is derived and applied in a systematic analysis on circuit level. Simulation and measurement results indicate, that moderate hysteresis effects are no show stopper for analog and mixed signal. Nevertheless exceptional cases have to be considered, corresponding countermeasures on circuit level are proposed and verified on silicon. The feasibility of important analog, mixed-signal and RF building blocks in an emerging multi-gate technology is proven by measurements, the performance is benchmarked against planar bulk. Multi-gate device specific design aspects are pointed out. Benefits on circuit level resulting from advantageous multi-gate device properties are explored: improved robustness and gain is demonstrated for current references and operational amplifiers. A significant reduction of circuit area is achieved for a 10 bit D/A converter. The use of gated p-i-n diodes in bandgap reference circuits is evaluated, a corresponding model covering temperature dependence is derived. Low-voltage bandgap references show competitive performance. Measured VCO and LNA characteristics indicate no road blocks for RF applications in the low GHz domain. Promising noise and jitter performance is demonstrated in a charge-pump PLL. Further multi-gate related design aspects like self-heating and selective tuning of fin width are outlined. Finally the integration of tunneling FETs in a low-power multi-gate technology is discussed as outlook to analog design aspects beyond CMOS. Although these devices feature low on-currents, promising analog properties and low variability regarding temperature and threshold voltage are demonstrated. Gate stack engineering and tuning of doping profiles are suited for device optimization. A TFET reference circuit is developed, robust against temperature and supply voltage variations.

Villach Michael Fulde

Danksagung

Die vorliegende Arbeit entstand während meiner Tätigkeit als wissenschaftlicher Mitarbeiter am Lehrstuhl für Technische Elektronik der Technischen Universität München. Für die Betreuung der Arbeit möchte ich Frau Prof. Doris Schmitt-Landsiedel herzlich danken. Ihr Engagement und Rat, sowie die Bereitschaft jederzeit und kurzfristig Probleme zu lösen trugen wesentlich zum Gelingen der Arbeit bei. Sehr großer Dank gilt allen ehemaligen Kollegen am Lehrstuhl für Technische Elektronik, die durch tatkräftige Hilfe, Motivation und ein angenehmes, freundschaftliches Arbeitsklima die Promotion sehr erleichtert haben. Auch bei meinen ehemaligen Diplomanden und Studenten möchte ich mich bedanken. Herrn Dr. Helmut Gräb danke ich für sein Interesse an der Thematik und für die Erstellung des Zweitgutachtens. Für die Unterstützung der Arbeit im Rahmen eines gemeinsamen Forschungs- und Entwicklungsvorhabens mit der Infineon Technologies Austria AG gilt der Dank Herrn Dr. Gerhard Knoblinger und dem gesamten ATE ACE Team. Ohne ihre hervorragende Unterstützung bei Design, Layout und Messtechnik wäre diese Arbeit nicht möglich gewesen. Ebenfalls danken möchte ich dem Münchner Infineon MuGFET Team um Klaus Schrüfer für wertvolle Diskussionen und tatkräftige Hilfe bei verschiedenen Messungen. Sehr großer Dank gilt den Kollegen am IMEC in Leuven, die im Rahmen des Nano-RF Projekts zahlreiche Messdaten, Modell-Parameter und natürlich die FinFET Hardware beigesteuert haben. Abschließend möchte ich mich bei meinen Eltern und meiner Freundin für ihr großartige Unterstützung und Geduld auch in hektischen Phasen der Promotion bedanken.

Contents

Chapter 1
Introduction

1.1 Motivation

Starting in the 1970s, information technology (IT) has drastically expanded until nowadays. The main applications pushing the enduring growth of IT and semiconductor industry are personal computers, network and communication technology including the Internet world and finally all kinds of portable consumer electronics. This success is mainly driven by the continuous scaling of CMOS as mainstream semiconductor technology, enabling efficient improvements in terms of processing speed, memory capacity, integration density and power consumption. Another aspect pushing the economic success of CMOS is the proceeding integration on a functional level, enabling further miniaturization and price reduction. Digital, memory, analog and mixed-signal blocks, even the radio-frequency (RF) front-end are integrated together in so called systems on chip (SOC) or systems in package (SIP). Especially in consumer electronics with high volume and high complexity there is a strong trend for SOC integration and single chip solutions. Semiconductor companies realized that in these business field SOC integration is an important differentiation factor and value adder. The use of the most recent CMOS technologies and smart design techniques is prerequisite for further improvements in terms of functionality, area and power consumption [1]. Especially portable electronics like wireless communication platforms have large analog and mixed-signal contents, e.g. voice processing, power management or RF interface. As illustrative example Fig. 1.1 shows the schematic view of a single chip mobile phone solution on block level. To obtain this high level of integration, the ability to realize analog, mixed-signal and RF circuits in scaled digital CMOS technologies with few specific analog and RF extensions is of outmost importance [2]. To meet future requirements in terms of bandwidth and power consumption in a cost effective manner, e.g. within new standards for mobile communication, further scaling of CMOS is needed also from analog and mixed-signal perspective [3, 4]. Consequently an early assessment of analog and mixed-signal circuit design feasibility in scaled digital CMOS is necessary.

Since device dimensions have reached the nanometer scale, conventional scaling concepts have been replaced, according to IBM's chief technology officer Meyerson

M. Fulde, *Variation Aware Analog and Mixed-Signal Circuit Design in Emerging Multi-Gate CMOS Technologies*, Springer Series in Advanced Microelectronics 28, DOI 10.1007/978-90-481-3280-5_1,

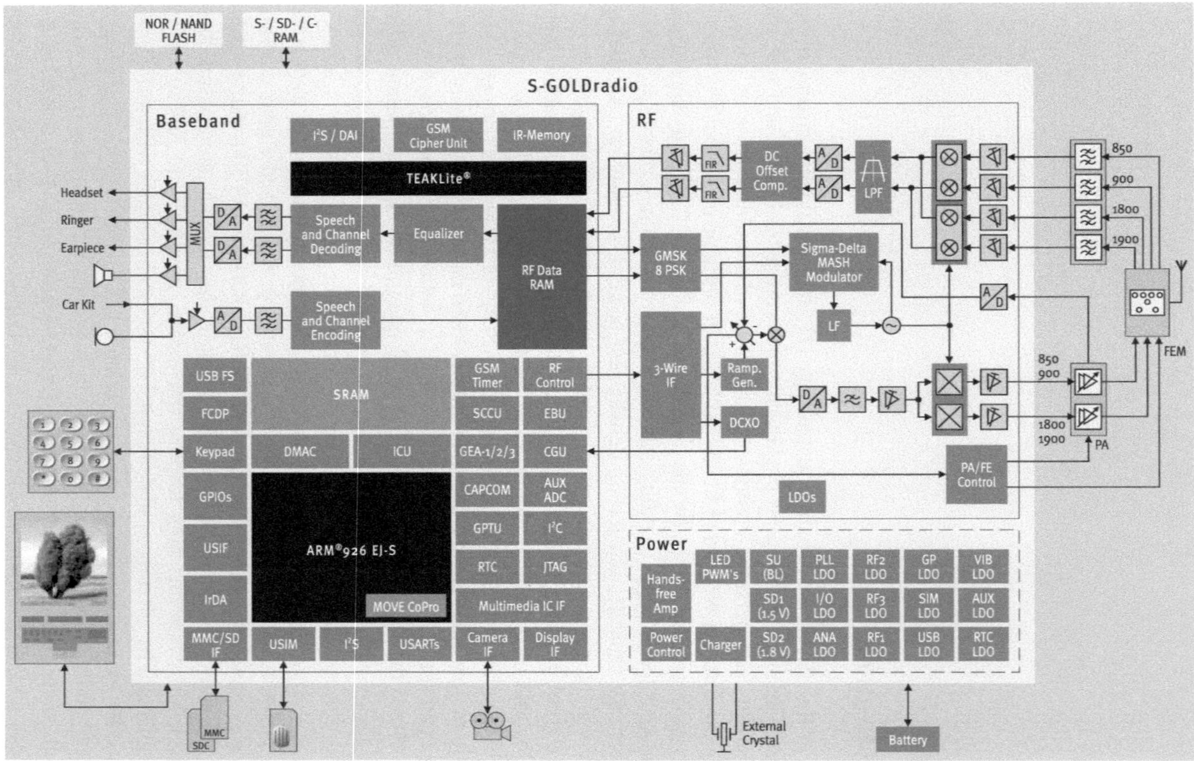

Fig. 1.1 Block diagram of S-GOLDradio^{TM}, a single chip mobile phone solution in 130 nm CMOS (source: http://www.infineon.com)

"traditional scaling died somewhere between the 130- and 90-nanometer nodes" [5]. In particular for sub 45 nm nodes revolutionary changes in technology are necessary to achieve reasonable transistor performance. Possible scaling scenarios include novel materials, novel device geometries and finally novel device concepts. Especially analog device performance and device variability is affected by these changes. As shown later in this chapter multi-gate transistors employing high-k gate dielectrics are considered as promising scenario overcoming the scaling limitations of conventional planar bulk CMOS. Benefits for multi-gate circuits in the digital domain have been already proven [6, 7]. Also basic analog functionality has already been shown in [8], however no conclusions on the performance scope are drawn.

Thus, the objective of this work is to provide a technology oriented, systematic and comprehensive overview of analog and mixed-signal circuit design aspects in emerging multi-gate technologies with focus on performance and variability aspects. The feasibility of analog and mixed-signal circuits in these technologies is assessed on device and circuit level. Multi-gate and high-k related aspects are separated, since advanced planar bulk CMOS including high-k dielectrics will be mainstream for the next 3–4 years [9]. As far as possible multi-gate is benchmarked against planar bulk. Key statements are supported by experimental results. The outline is as follows:

In the remaining part of Chap. 1 the fundamentals of non-classical scaling beyond the constant-field approach are shown. Different scaling options suitable for near-, mid- and long term requirements are outlined. In addition the most important variation effects are classified and related to analog and mixed-signal circuit performance. Chapter 2 introduces state-of-the-art multi-gate technology with respect to analog device behavior and variability, covering also the impact of high-k dielectrics. Analog and mixed-signal circuit design issues related to high-k are analyzed in Chap. 3. The impact of increased flicker noise and dynamic variations of threshold voltage are discussed exemplary. Countermeasures on circuit level are sketched. The introduction of multi-gate MOSFETs (MuGFETs) as successor of planar bulk devices represents fundamental changes in device structure. Chapter 4 analyzes the consequences on circuit level. The feasibility of analog, mixed-signal and RF building blocks is assessed. Benefits and limitations corresponding to the multi-gate structure are identified. The multi-gate tunneling FET (MuGTFET) is investigated as example for an alternative device concept in Chap. 5. Analog design considerations are derived from basic device performance, temperature and matching behavior. A new voltage reference circuit is proposed making use of the specific tunneling device characteristics. Chapter 6 gives the final conclusion and future prospects.

1.2 Scaling Fundamentals

The great scalability of CMOS results from the field-effect principle of operation. The transistor behavior is determined by the electric fields within the device. Scaling the geometrical device dimensions such as gate length (L_{gate}), junction depth (x_j)

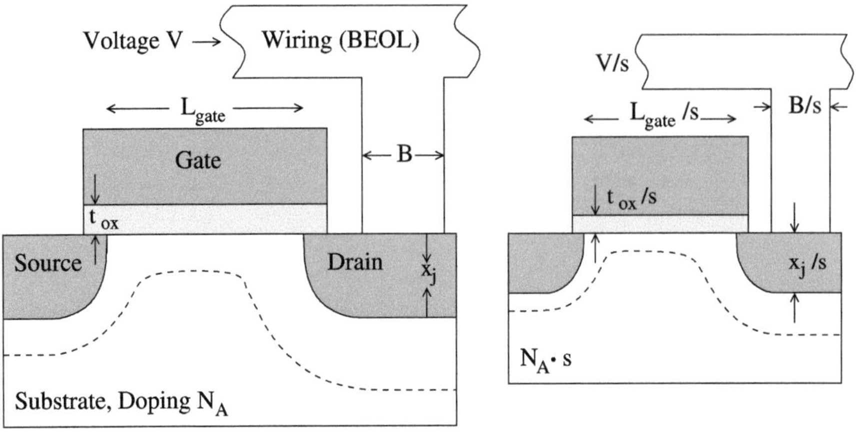

Parameter		Scaling
Electric Field	F	1
Transistor Delay	τ	1/s
Area	A	$1/s^2$
Power per Operation	P	$1/s^3$

Fig. 1.2 Schematic view of MOSFET and table of important scaling parameters

and oxide thickness (t_{ox}), the channel doping[1] (N_A) and the supply voltage (V_{DD}) with a constant factor s does not affect the electric fields in ideal case, see Fig. 1.2. However, the transistor delay is reduced by the same factor s, the area by the factor s^2 and the power consumption per operation even by s^3. Typically a scaling factor of about 1.4 has been used from node to node, yielding an area shrink factor of 2.

This constant-field scaling has been a key factor for technical and economic success of CMOS technology for many years. To maintain this trend a consortium of major semiconductor companies specifies scaling targets, challenges and scenarios in the "International Technology Roadmap for Semiconductors".[2]

Since device dimensions reached the nanometer regime a couple of years ago, the traditional constant field scaling is no longer applicable. Previously negligible phenomena like quantum mechanical tunneling currents or short-channel effects now significantly impact transistor behavior as classical planar bulk CMOS technology is approaching its physical limits. The main consequences are increasing leakage currents, increasing variability and degrading analog transistor performance, enforcing the introduction of novel concepts.

[1] N_A is not downscaled but increased by s.

[2] http://www.itrs.net

1.2.1 New Materials: High-k Gate Dielectrics

Since the first MOSFETs have been fabricated on silicon over 40 years ago, SiO_2 has been used as gate dielectric due to its straightforward manufacturing and beneficial electrical and thermal properties. According to constant-field scaling the oxide thickness was scaled down until 2–4 nm at the 130 nm (high-performance) or 90 nm (low-power) node, where the dielectric film comprises only few atomic layers. Below a physical thickness of 3 nm quantum-mechanical mechanisms like direct tunneling from gate into the channel region contribute to severe gate leakage current issues.

To overcome this problem extensive research has been spent on insulating materials with higher dielectric constant κ $(=\epsilon_r)$[3] than SiO_2 ($\kappa = 3.9$). Higher κ values allow to scale down the electrical (= equivalent) oxide thickness (EOT) according to

$$\mathrm{EOT} = t_{ox}\frac{\kappa_{\mathrm{SiO_2}}}{\kappa}, \tag{1.1}$$

whereas the physical thickness t_{ox} remains constant or is even increased. The use of nitrided silicon oxide SiON ($\kappa = 4 \ldots 6$) as "slightly higher-k" material enabled acceptable gate leakage in 90 nm [10], 65 nm [11, 12] and 45 nm [13, 14] nodes as straightforward but limited extension of SiO_2. The main challenges in the integration of a "real" high-k material like hafnium oxide HfO_2 ($\kappa \approx 25$) result from the fact, that the dielectric can not be thermally grown on the silicon surface but has to be deposited [15, 16]. Sufficient interface quality is required to ensure reasonable electrical and thermal stability, mobility of charge carriers, flicker-noise and matching performance.

Beyond that other difficulties in replacing SiO_2 with high-k materials have been reported in literature, arising from the interaction of the poly-silicon gate electrode with the high-k dielectric. Fermi-level pinning causes too high threshold voltage V_T [17], surface phonon scattering severely degrades channel electron mobility [18]. To circumvent these issues the conventional poly gates have to be replaced by metal gate electrodes consisting e.g. of TiN or TaN [18]. Since the threshold-voltage of metal gate transistors is determined by the workfunction of the metal, the main challenge here is to find proper material systems with suitable workfunctions for V_T adjustment [19].

Intel as the major manufacturer of microprocessors successfully introduced a high-k, metal gate process with its 45 nm high-performance technology [20]. However, with respect to the requirements for high-resolution and high-speed analog and mixed-signal systems there are still open challenges on technology and circuit level, as will be shown in Chap. 2.

Although the introduction of high-k dielectrics and metal gates seems to solve gate leakage problems and to enable further scaling, planar bulk CMOS will face significant challenges going beyond 32 nm: The doping of the channel is reaching

[3] Although the symbol for the dielectric constant is ϵ_r or κ the term high-k has become popular.

undesirably high levels in order to gain adequate control of short-channel effects and to set the threshold voltage properly. As a result of the high channel doping, the mobility of holes and electrons is reduced and the junction leakage due to band-to-band tunneling and gate-induced drain leakage will increase. Furthermore, statistical variations of the threshold voltage are increased due to the small total number of dopants in the channel of extremely small MOSFETs. Consequently the next revolutionary step in CMOS scaling will be a change of the device geometry.

1.2.2 New Device Architectures: Multi-Gate MOSFETs

With scaled transistor dimensions, the close proximity of source and drain region reduces the ability of the gate electrode to control the potential distribution and current flow in the channel region. There are two main so called short-channel effects (SCE) that will limit the scaling of conventional planar bulk MOSFETs somewhere around the 22 nm node [9]: the dependency of the threshold voltage on the gate length, i.e. on the distance between source and drain, called V_T roll-off (VTRO) and the dependency of V_T on the drain bias, called drain induced barrier lowering (DIBL). Both effects lead to drastically increased subthreshold leakage currents. Taking SCE into account the threshold voltage of a MOSFET can be written as

$$V_T = V_{T0} - \text{VTRO} - \text{DIBL}, \tag{1.2}$$

where V_{T0} is the threshold voltage of an ideal long channel device. Using the voltage-doping transformation model [21] the short-channel effects VTRO and DIBL can be related to device geometry and bias parameters [22]:

$$\text{VTRO} = 0.64\frac{\kappa_{si}}{\kappa_{ox}}\text{EI} \cdot V_{bi} \tag{1.3}$$

and

$$\text{DIBL} = 0.84\frac{\kappa_{si}}{\kappa_{ox}}\text{EI} \cdot V_{ds} \tag{1.4}$$

with the dielectric constants of silicon κ_{si} and gate oxide κ_{ox}, the built-in source potential V_{bi} and the electrostatic-integrity factor EI. The electrostatic-integrity is a measure describing the way the electric field from the drain influences the channel region and depends on device geometry. For a planar bulk MOSFET as shown in Fig. 1.3, the electrostatic-integrity is given as

$$\text{EI} = \left(1 + \frac{x_j^2}{L_{el}^2}\right)\frac{t_{ox}}{L_{el}}\frac{t_{dep}}{L_{el}}. \tag{1.5}$$

L_{el} is the effective channel length and t_{dep} the penetration depth of the gate field in the channel region. It is obvious that short-channel effects can be minimized scaling

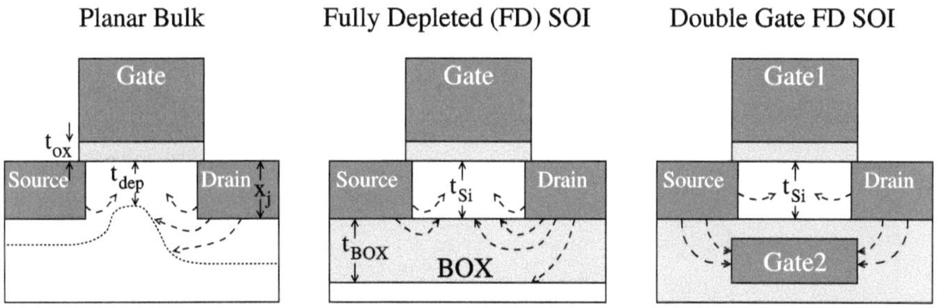

$$\mathrm{EI} = \left(1 + \frac{x_j^2}{L_{el}^2}\right) \frac{t_{ox}}{L_{el}} \frac{t_{dep}}{L_{el}} \quad \mathrm{EI} = \left(1 + \frac{t_{si}^2}{L_{el}^2}\right) \frac{t_{ox}}{L_{el}} \frac{t_{si}+\lambda t_{BOX}}{L_{el}} \quad \mathrm{EI} = \left(1 + \frac{t_{si}^2}{4L_{el}^2}\right) \frac{t_{ox}}{L_{el}} \frac{t_{si}}{2L_{el}}$$

Fig. 1.3 Propagation of electric-field lines and electrostatic integrity for planar bulk (*left*), fully-depleted SOI (*middle*) and fully-depleted double gate SOI MOSFET (*right*)

down x_j, t_{ox} or reducing t_{dep} by increasing the channel doping. As mentioned above there are some practical limitations for these measures.

Short-channel effects depend on the way the electric field lines from source and drain affect the gate control on the channel region. In planar bulk devices the field lines propagate through the depletion zones as shown in Fig. 1.3. The situation is different in fully depleted (FD) silicon on insulator (SOI) devices. These transistors are built in a thin silicon layer on top of a buried oxide layer (BOX). The thickness of the silicon layer t_{si} is small enough to be fully depleted by the gate field. In this case most of the field lines propagate through the BOX. As a consequence the electrostatic-integrity and the short-channel behavior of FD-SOI MOSFETs strongly depends on the thickness of the silicon and oxide layers, see Fig. 1.3. Gate lengths between 20 nm and 10 nm seem possible with FD-SOI MOSFETs featuring very thin silicon layers in the range of few nanometers [9]. Since the electrical behavior of these transistors is strongly dependent on t_{si}, the film thickness has to be controlled very accurately over the whole wafer to keep device parameter variations under control. However, the manufacturing of such thin layers with sufficient conformity is quite challenging [23]. The tight requirements on film thickness are considerably relaxed, if a second gate is used to improve the channel control and electrostatic-integrity [24]. In this kind of double gate devices both gates are connected together, in contrast to independent double gate FETs [25]. The field lines below the channel region terminate at the bottom gate electrode and therefore have no impact on the short-channel behavior of the device, as illustrated in Fig. 1.3. In a double gate device the effective junction depth and gate field penetration for each gate is $t_{si}/2$. Roughly speaking the double gate MOSFET looks twice as thin as the equivalent FD-SOI MOSFET from an electrostatic-integrity point of view, short-channel effects are significantly reduced.

However, manufacturing a self-aligned double gate MOSFET is anything but straightforward. Five years after it was proposed, the first self-aligned vertical multi-gate device, called DELTA (fully DEpleted Lean channel TrAnsistor) was fabricated in 1989 [26]. In this device the gate is wrapped around a thin vertical silicon island,

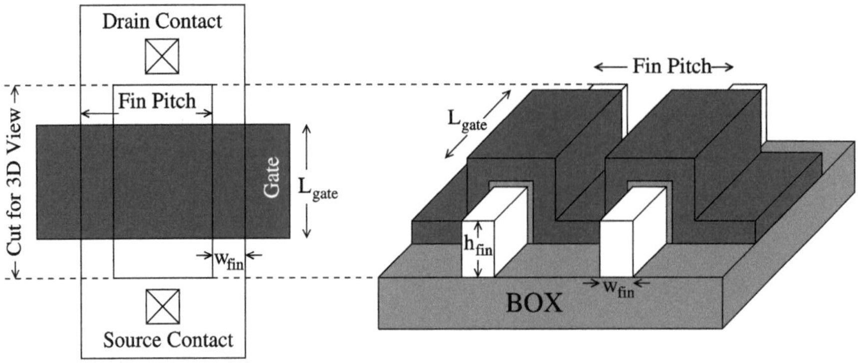

Fig. 1.4 Top and 3-dimensional view of a FinFET with two fins in parallel. Source and drain contact landing pads are indicated *on the left* and cutted in the 3D view

called fin, on top of an oxide layer. Due to its compatibility to planar CMOS processing this structure became the basis for almost all further multi-gate structures. Until now a whole zoo of multi-gate devices has been proposed in literature differing in the form of the silicon island and the numbers of effective gate electrodes around it, e.g. FinFET [27], Tri-Gate FET [28], Pi-Gate FET [29], or Omega FET [30].

For the sake of clearness in the following a simplified terminology is introduced: multi-gate MOSFET (MuGFET) is the generic term for devices consisting of a thin wire or fin like silicon island surrounded by a gate. The term FinFET is a subclass of these transistors, consisting of one or more parallel silicon fins with the gate stack wrap around, controlling the channels on the fin sidewalls and optional on top of the fin, see Fig. 1.4. This work covers only FinFETs, since they are currently seen as most promising multi-gate device structure to replace planar bulk.

Obviously the third gate electrode on top of the fin improves the electrostatic-integrity further. Compared to a double gate device the requirements on the silicon film thickness, i.e. the fin width w_{fin} are reduced. A detailed analysis of the electrostatic dependencies on device geometries can be found in [31]. Figure 1.4 shows a top and 3-dimensional view of a FinFET including important geometrical parameters. It is worth noting that the minimum feature size of a FinFET is the fin width, not the gate length. A rule of thumb to keep short-channel effects under control is to choose the ratio of gate length to fin width not below $3/2$ [32]. A recent example for a low-power multi-gate CMOS technology is shown in Chap. 2.

1.2.3 New Device Concepts: Tunneling FETs

Although dimensional scaling of CMOS can be extended with multi-gate devices to gate lengths in the range of 10 nm and below, there is certainly a need for alternative devices beyond, able to overcome the MOSFET scaling limitations. The idea is to change the principle of device operation itself rather than the device structure.

Different new emerging technologies for memory and information processing have already been proposed [33]. A second motivation for an early analog assessment of novel device concepts is the heterogeneous integration of these new technologies with CMOS platforms, called "enhanced CMOS" or "more than Moore", enabling further functional diversification. Promising concepts mentioned in the ITRS roadmap are one-dimensional structures like carbon nanotube FETs, single electron and spin transistors, molecular devices, ferromagnetic logic and interband tunneling transistors [33].

In this work the Tunneling Field Effect Transistor (TFET) serves as example for a promising novel device concept. The TFET principle of operation is based on a gate controlled Zener tunneling current [34]. The characteristics of this device is determined by the narrow tunneling junction and therefore insensitive to variations of the channel length. Device simulations show that the spatial expansion of tunneling junctions is smaller than 10 nm which proves the scaling potential of the TFET [35]. Lateral and vertical tunneling FETs have already been presented in literature [36, 37]. Combining the scaling potential of three-dimensional and tunneling devices, in this work complementary TFETs have been realized in a SOI multi-gate FET technology.

In Chap. 5 the specific integration of tunneling FETs in a state-of-the-art multi-gate CMOS technology is shown. Device characteristics are analyzed form analog point of view, temperature and matching behavior are discussed. Analog design considerations with special regard to variability are also given.

1.3 Variability from Analog and Mixed-Signal Perspective

Summarizing the previous sections the key statements are: SOC designs require high resolution and robust analog and mixed-signal circuits in scaled digital CMOS and these technologies will significantly differ from conventional CMOS: for sub 32 nm nodes multi-gate devices with high-k/metal gate are an attractive solution.

The assessment of analog feasibility is based on performance and variability in general, since all key analog and mixed-signal circuit performance metrics like linearity, resolution or power consumption are determined by different kinds of variation effects. These limitations are derived in [38] for device parameter mismatch and will be extended to other effects in this work. This section classifies the most important variation effects on device and circuit level.

1.3.1 Systematic Static Variations

In this work the term systematic static variation denotes a circuit non-ideality caused by imperfect MOSFET behavior in a deterministic and static sense. That means a certain quantity (e.g. the output voltage) is affected by disturbances or harmonic distortion due to non-ideal MOSFET current sources. Although the output signal

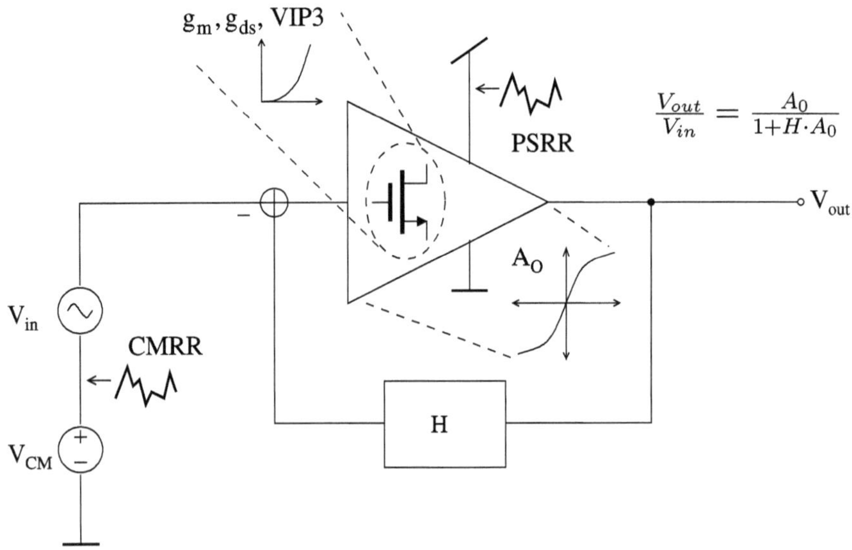

Fig. 1.5 Systematic static variations in amplifiers with feedback

variations are visible in the time or frequency domain, the root-cause is related to systematic (i.e. deterministic), time-invariant transistor quantities. Typical analog circuits suffering from systematic static variations are operational amplifiers in feedback configuration, intended to deliver an input signal V_{in} with a specified transfer function to a load, e.g. as buffer, amplifier or integrator, see Fig. 1.5. In this kind of analog circuits MOSFETs are mostly used as constant or controlled current sources. Possible systematic current source imperfections are finite output resistance, i.e. non-zero output conductance g_{ds}, limited intrinsic gain g_m/g_{ds} and non-linear transfer characteristics. In analog building blocks like references or amplifiers a finite current source output conductance typically translates in sensitivity against variations of common mode level V_{CM} or supply voltage V_{DD}, quantitatively specified as Common Mode Rejection Ratio (CMRR) and Power Supply Rejection Ratio (PSRR).

The intrinsic transistor gain g_m/g_{ds} represents the maximum voltage gain of a single transistor. This gain limits the DC open-loop gain of operational amplifiers A_0, see Chap. 4. Since the absolute value of DC gain is typically not well defined, operational amplifiers (OpAmps) are commonly used in negative feedback configurations to generate a signal transfer function with high accuracy or resolution [39]. As long as the open-loop gain of the OpAmp is sufficiently high, the absolute value of the closed-loop signal transfer function is only determined by the well defined passive feedback network. The feedback also reduces the impact of (non-linear) open-loop gain variations, i.e. the generation of harmonic distortion is suppressed. Roughly speaking, limited g_m/g_{ds} results in finite OpAmp gain and limited accuracy and resolution.

Another source of harmonic distortion is the non-linear transfer characteristics of the MOSFET itself. The MOSFET linearity is commonly characterized by the third order interception point VIP3, defined as the value of gate source voltage V_{GS} where the drain current of the fundamental tone equals that of the third harmonic [40]:

$$\text{VIP3} = \sqrt{24\left|\frac{g_m}{g_{m3}}\right|}; \quad g_{m3} = \frac{\partial^3 I_D}{\partial V_{GS}^3}. \tag{1.6}$$

The requirements for high sensitivity against systematic static variations on device level are low g_{ds}, high g_m/g_{ds} and high VIP3. On circuit level these parameters correspond to high open loop gain, high PSRR, high CMRR and low harmonic distortion.

1.3.2 Static Random Variations—Mismatch

Process induced random fluctuations of transistor parameters like V_T, μC_{ox} or W/L cause a mismatch of nominal identical devices. The physical origin of mismatch includes random fluctuation of channel and gate doping, random fluctuations in surface roughness and oxide charge and finally lithography induced line edge roughness. In today's CMOS technologies typically the random fluctuation of channel doping resulting in a V_T variation is the main source of mismatch. A simple model translates the variation of device parameters into drain current variations by

$$\sigma_{\Delta I_D/I_D} = \frac{g_m}{I_D}\sigma_{\Delta V_T} + \beta\sigma_{\Delta\beta/\beta_0} \tag{1.7}$$

with $\beta = \frac{W}{L}\mu C_{ox}$ [41]. The geometrical dependence of the device parameter mismatch can be described as

$$\sigma_{\Delta V_T} = \frac{A_{V_T}}{\sqrt{WL}}; \qquad \sigma_{\Delta\beta/\beta_0} = \frac{A_\beta}{\sqrt{WL}} \tag{1.8}$$

with the process dependent matching constants A_{V_T} and A_β [42]. Empirical observations show that A_{V_T} scales with channel doping N_A and oxide thickness t_{ox} [43]

$$A_{V_T} \propto t_{ox} \cdot \sqrt[\alpha]{N_A} \tag{1.9}$$

where α is close to 2. Following the CMOS scaling rules for N_A and t_{ox} in the past A_{V_T} scaled down with technology. However, different mechanisms have started to limit the scaling of A_{V_T} recently [44]: as shown above gate leakage limitations prevent further EOT scaling and other mismatch effects such as line edge roughness gain in importance as optical lithography faces its physical limits.

From analog and mixed-signal circuit perspective stochastic device parameter mismatch generally limits accuracy and resolution. To improve device matching, increased device area and consequently increased power consumption has to be accepted. Assuming V_T mismatch to be dominant, it can be shown that the speed-accuracy-power trade-off in analog CMOS circuits is determined by the V_T matching constant [38]:

$$\frac{\text{Bandwidth} \cdot \text{Accuracy}^2}{\text{Power}} \propto \frac{1}{A_{V_T}^2}. \tag{1.10}$$

1.3.3 Transient Random Variations—Noise

Besides static device parameter variations of course dynamic effects have to be considered. Small dynamic random fluctuations of voltages or currents are called electrical noise. Only noise generated within the devices themselves, i.e. no cross-talk etc. is considered here. The existence of noise is basically due to thermal energy and the fact that electrical charge is not continuous but is carried in discrete amounts. The main sources of noise in MOSFETs are shot noise, burst or random telegraph noise, thermal noise and flicker noise. Due to the capacitive coupling any drain current noise yields also gate current noise, which has to be considered at high frequencies. In typical CMOS analog circuits thermal noise and flicker noise are dominating [45]. Since thermal noise is present in every electrical conductor and in first order independent of technology this work is focused on flicker noise. Flicker noise is generated in the interface region of silicon-channel and gate dielectric layer(s). Interface states and oxide traps capture and release charge carriers by direct tunneling in a random process. The superposition of the different time constants corresponding to the statistically distributed traps typically yields a $1/f$ frequency dependence. Hence flicker noise is also called $1/f$ noise. Obviously the amount of flicker noise heavily depends on technology, respectively on the silicon orientation, the silicon-dielectric interfacial layer, the dielectric material itself and other process parameters [46, 47]. A simple model for the spectral flicker noise density related to drain current and gate voltage is given in [45]:

$$\frac{\overline{i_{1/f}^2}}{\Delta f} = \frac{K_f I_D}{C_{ox} L^2 f^{ef}}, \tag{1.11}$$

$$\frac{\overline{v_{1/f}^2}}{\Delta f} = \frac{K_f}{2\mu C_{ox} W L f^{ef}} \tag{1.12}$$

with the fitting parameters $K_f \approx 10^{-28} \ldots 10^{-26}$ $\text{A}^2\text{s/V}$ and $ef \approx 1$.

In circuits noise represents a lower limit to the signal level that can be processed. Similar to mismatch noise improvement is paid with area. Consequently noise limits its accuracy, speed, area, and/or minimal power dissipation. Thus, sufficient low (flicker) noise is a strong requirement for any new CMOS technology from analog point of view. Especially the introduction of high-k dielectrics is challenging in terms of flicker noise, since no thermal oxide is "simply" grown on the silicon crystal with inherently good interface quality but the dielectric has to be deposited on the silicon surface.

1.3.4 Transient Systematic Variations

Deterministic, time varying device parameters are well known in CMOS technologies, particularly in relation to device lifetime and reliability. Bias-temperature instabilities and hot-carrier degradation significantly shift device characteristics, but on long time scales and high stress voltages, far away from typical analog operating conditions. However, in emerging CMOS technologies also deterministic dynamic variations of device behavior in relevant operating conditions have to be considered. Charge-trapping in high-k dielectrics and self-heating in SOI devices are considered in this work.

Trapping and de-trapping of charge carriers in pre-existing states and defects is associated to flicker noise above. Under large signal conditions not only random current fluctuations but also deterministic transient threshold voltage shifts are observed [48]. The filling of traps is induced by the gate field, resulting in a dynamic V_T shift. Similar to flicker noise charge-trapping is very sensitive to technology, respectively to the gate stack and its interface to the silicon [49]. The reason is the identical physical origin of both effects. Thus, high-k dielectrics are expected to suffer from enhanced charge-trapping.

The trapping time constants range from sub-micro to milliseconds. Without gate bias V_T typically fully recovers its pre-stress value. Figure 1.6(a) illustrates the time dependence of I_D and V_T shift for a rectangular signal at the transistor gate. With rising gate bias, the current follows until reaching a peak value. Also V_T increases

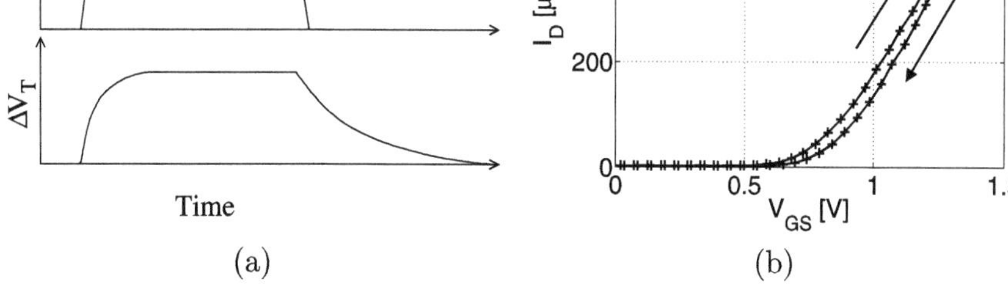

Fig. 1.6 Schematic time dependence of I_D and ΔV_T for gate pulse (**a**) and corresponding measured I_D–V_{GS} hysteresis curve (**b**)

with a certain time constant due to charge trapping which results in a decay of drain current until a steady state is reached. After turning-off the gate bias, V_T recovers to its pre-stress value. A corresponding I_D–V_{GS} characteristics is shown in Fig. 1.6(b). The measured planar bulk nFET with HfO_2 high-k dielectric and poly-Si gate exhibits pronounced hysteresis behavior, the corresponding V_T shift amounts to several 10 mV in this case. It has to be noticed that these values are obtained from an experimental process as illustrative example. State-of-the-art high-k/metal gate stacks typically show V_T shifts in the range of some mV at 1 V stress voltage. The consequence of this effect on circuit level is a history dependent device behavior [50], similar to floating body effects in partially depleted SOI devices.

Another source of dynamic device variation is self heating: the power dissipation of an active device causes a local temperature increase [51], changing the actual transistor current due to the temperature dependence of device parameters such as V_T or μ. The time-constants of this effect are typically in the nanosecond regime, close to operating frequencies of high-speed analog and mixed-signal circuits. Devices on SOI substrates are much more affected by self-heating due to the poor thermal conductivity of the surrounding oxide [52, 53]. Especially small volume devices like fully depleted thin body FETs or FinFETs with high power density severely suffer from self-heating.

Both effects cause a history dependent transistor behavior. Consequently devices show transient mismatch in case of different operating history. Again accuracy and resolution of analog and mixed-signal circuits is limited. However in contrast to mismatch and noise there is no "simple" trade-off in resolution versus area and power. Particular countermeasures on layout, circuit or architectural level are required, shown in Chaps. 3 and 4.

Chapter 2
Analog Properties of Multi-Gate MOSFETs

This chapter reviews the main multi-gate device characteristics relevant for analog and mixed-signal circuit design. The objective is to close the link from technology and integration aspects to analog device performance. The associated trade-offs are outlined. The introduction of three-dimensional multi-gate devices with high-k/metal gate stack represents "revolutionary" changes in CMOS technology. Therefore a close insight to analog device behavior serves as basis for the technology oriented circuit design issues addressed later.

First an example for a recent FinFET technology is introduced. Then DC, small signal, RF, noise and matching performance is discussed and compared to similar planar devices. Finally charge-trapping and self-heating are covered as example for novel technology related device effects.

2.1 Introduction to Recent FinFET Technology

All measurement and simulation results shown here are based on recent low-power multi-gate CMOS technologies as presented in [6] and [54]. Technology details for the planar reference devices are given in [55] and [56].

The basic FinFET process flow is illustrated in Fig. 2.1. Fins are patterned in a 60 nm high undoped[1] silicon layer on standard 200 mm or 300 mm SOI wafers. Dedicated optical proximity correction (OPC) allows to process fin widths w_{fin} down to 10 nm. For circuit evaluation in this work w_{fin} typically is chosen around 20 nm...30 nm. The effective channel width per fin in these triple gate devices is $2h_{fin} + w_{fin} \approx 0.14$ µm. Taking the fin pitch of 200 nm into account, the area efficiency is reduced compared to planar. However, several techniques to reduce the fin pitch and improve the area efficiency have already been proposed, e.g. double patterning [57]. HfSiON is deposited as high-k dielectric, yielding EOT values of about 1.6 nm. For comparison also SiON reference devices with EOT of 1.9 nm

[1] Undoped means doping levels below 10^{15} cm^{-3}.

M. Fulde, *Variation Aware Analog and Mixed-Signal Circuit Design in Emerging Multi-Gate CMOS Technologies*, Springer Series in Advanced Microelectronics 28,
DOI 10.1007/978-90-481-3280-5_2,

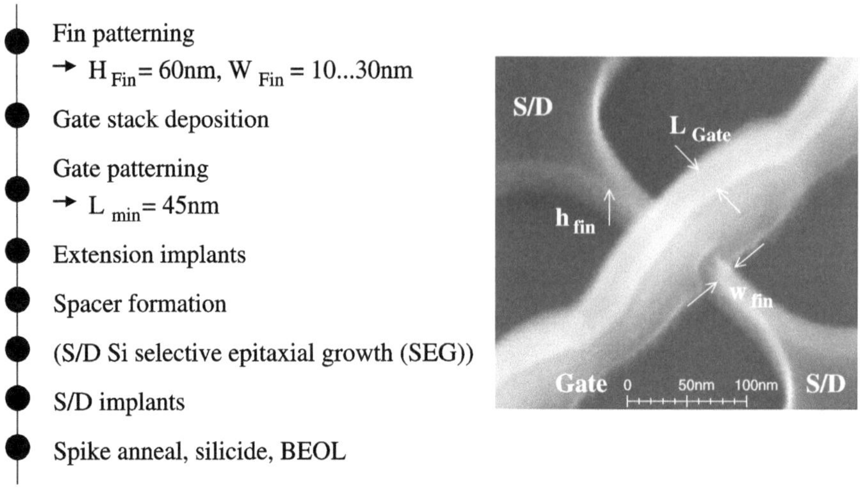

Fig. 2.1 FinFET process flow and SEM picture of single fin device after gate patterning

are processed. TiN is used as gate electrode material. Due to the undoped fins mid-gap workfunction is required for symmetric nFET and pFET threshold voltages. To achieve mid-gap workfunction on both dielectrics the TiN thickness is adjusted between 5 nm and 10 nm. If needed, additional capping and V_T adjust layer(s) are deposited below and on top of the metal gate. The gate stack is completed with a poly silicon layer which is finally silicided to reduce the contact resistance. A concrete example for a high-k/metal gate stack is shown later in this chapter. Additional gate etching enables nominal gate lengths of 45 nm [54] and 65 nm [6]. After spacer formation the fin width is optionally increased by selective epitaxial growth. Then source/drain regions are implanted vertically or 45° tilted. Further process steps include annealing, silicidation and finally copper metallization (BEOL).

2.2 DC Characteristics

It was shown in Chap. 1 that the electrostatic integrity, i.e. the short-channel behavior is determined by the relation of fin thickness to channel length at given EOT. A rule of thumb to keep short channel effects under control is that the fin width w_{fin} should not exceed 2/3 of the minimum channel length [31, 32]. For the devices considered here this requirements is always fulfilled. The measured transfer characteristics of a n-type FinFET featuring minimum gate length $L_{min} = 45$ nm shown in Fig. 2.2 proves that excellent short channel behavior is achievable with proper engineered FinFETs. Very low DIBL effect (<25 mV) and close to ideal sub-threshold slope ($\approx$70 mV/Dec) are obtained, see Table 2.1. The well controlled short channel effects enable very low leakage currents (measured at $V_{GS} = 0$ V and $V_{DS} = V_{DD} = 1.0$ V), whereas the on-currents (measured at $V_{GS} = V_{DS} = V_{DD}$)

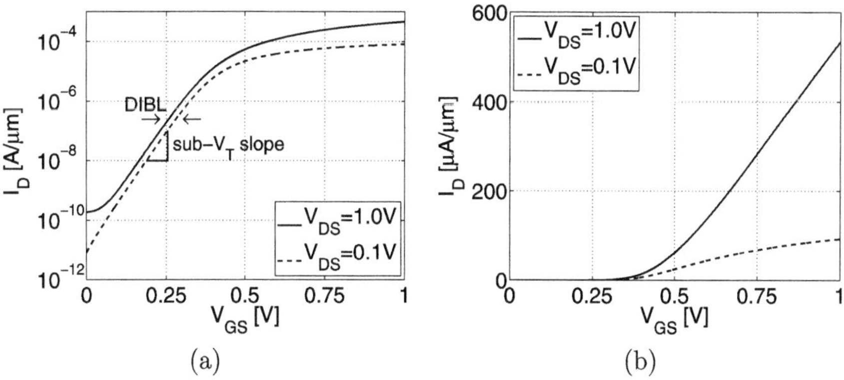

Fig. 2.2 Measured transfer characteristic in saturation (*full lines*) and linear (*dashed lines*) regime of n-type FinFET with $L_{gate} = 45$ nm

Table 2.1 Digital performance of FinFET with $L_{gate} = 45$ nm and $V_{DD} = 1.0$ V

	nFET	pFET
I_{ON} [µA]	534	452
I_{OFF} [nA]	0.31	0.42
I_{GATE} [nA]	0.09	0.07
V_T [V]	0.35	−0.34
DIBL [mV]	24	26
Sub-V_T slope [mV/Dec]	69	73

Fig. 2.3 Measured output characteristic of n-type FinFET with $L_{gate} = 45$ nm for V_{GS} from 0.2 to 1.0 V

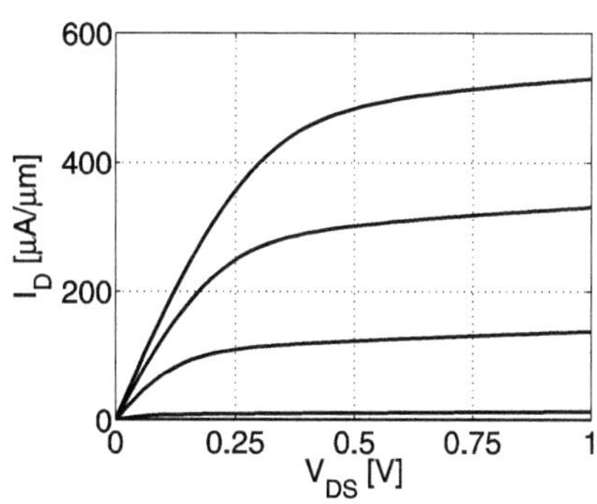

meet the ITRS targets for low standby power devices. The off-currents in this example are not limited by sub-threshold leakage but by gate-induced drain leakage. Gate leakage currents are suppressed by the high-k dielectric. The digital device characteristics is summarized in Table 2.1, nFET and pFET feature almost the same performance. The high pFET on-current results from the (110) surface orientation of the fin sidewalls with high hole mobility. The reduced short channel effects affect also the output characteristics I_D–V_{DS} shown in Fig. 2.3: in contrast to aggressively

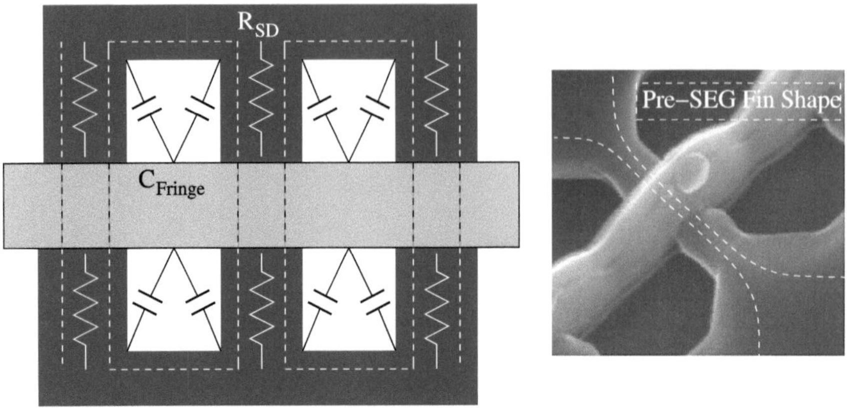

Fig. 2.4 Schematic view and SEM picture of FinFET before and after SEG including parasitics

scaled planar bulk CMOS technologies [58] the slope of the drain current in saturation is not increased by pronounced DIBL effect.

Another FinFET specific characteristic is visible in the linear I_D–V_{GS} plot of Fig. 2.2(b). The slope of the drain current in saturation flattens form quadratic to linear behavior at lower V_{GS} values than usually observed in planar bulk devices. This effect can be explained by high parasitic source/drain series resistances R_{SD} in the narrow fins which connect the channel to the source/drain contact landing pads, see Fig. 2.1.

Wider fins are no viable solution for this issue because short-channel and leakage requirements enforce a certain L_{gate}/w_{fin} ratio. Besides different mobility enhancement techniques one approach to improve the drive current therefore is to widen the fins outside of the gate, e.g. by selective epitaxial growth (SEG). SEG as illustrated in Fig. 2.4 has been used to decrease the source/drain resistances yielding higher drive currents [59]. However, the widening of the fins results in increased parasitic fringe capacitances which affect the RF performance. Nevertheless, it has been shown recently that despite of the additional parasitic capacitances FinFETs can achieve performance advantages over planar bulk at 22 nm node dimensions [60].

2.3 Analog and RF Characteristics

The most important analog and RF device figures-of-merit determining circuit resolution, speed and power consumption are the small signal parameters, noise and matching behavior. As starting point for analog and RF circuit design typically a simple linear equivalent circuit model is used [61]. The transistor behavior is linearized in a certain bias point and modeled by linear sources, resistors and capacitances, for more details see [62]. The slopes in the linearized approximation are called small signal parameters.

2.3.1 Small Signal Parameters

The relevant parameters regarding the analog performance are the transconductance g_m, the output conductance g_{ds} and the intrinsic gain g_m/g_{ds}. The transconductance describes how efficient a small voltage signal at the transistor gate is converted into a drain current signal. On circuit level the available g_m limits e.g. the bandwidth of operational amplifiers. Figure 2.5(a) shows measured gate length dependence of g_m for n-type FinFET and planar reference devices at typical analog bias conditions ($V_{GS} = V_T + 200$ mV and $V_{DS} = 1$ V). The FinFET g_m is slightly lower compared to the planar reference mainly due to the high parasitic source/drain resistances [63]. The situation gets worse for shorter channel lengths and increasing overdrive voltage $V_{GS} - V_T$.

The output conductance g_{ds} describes the quality of the device as constant current source. In modern planar CMOS technologies g_{ds} is not only determined by the channel length modulation but furthermore degraded for different reasons: pronounced DIBL effects cause a significant V_T dependence on V_{DS} which again leads to an enhanced (non-linear) drain current dependence on V_{DS}, visible as degraded output conductance. In order to keep short-channel-effects at tolerable levels typically strong pocket or HALO implants are necessary in planar bulk devices. These implants close to the source/drain extensions are intended to confine the source/drain field penetration. Implicitly an additional barrier is introduced at the drain end of the channel which can be modulated with the drain source bias. Especially at longer gate lengths this barrier results in a large residual DIBL-like effect yielding again degraded g_{ds} values [64]. The effect is sometimes called drain induced threshold shift to distinguish from DIBL because the effect is related to the drain barrier. This g_{ds} degradation for long channel devices strongly affects analog circuit design in scaled CMOS, since the efficiency of the common way to enhance output resistance is partly lost.

In fully-depleted FinFET devices the main contributors to the g_{ds} degradation are no concern: due to the enhanced electrostatic integrity short channel effects

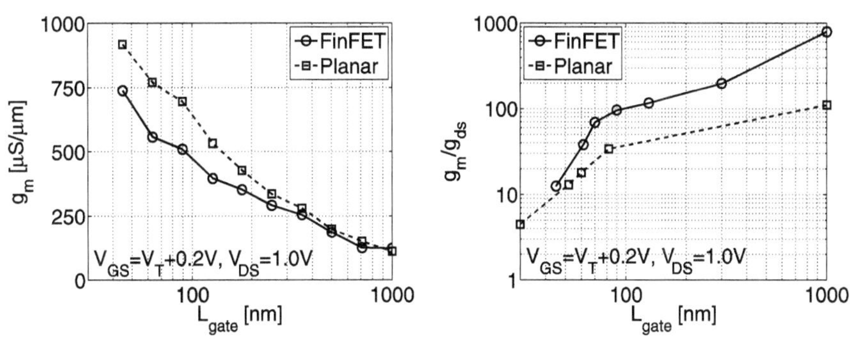

(a) (b)

Fig. 2.5 Measured g_m (**a**) and intrinsic gain g_m/g_{ds} (**b**) of n-type planar and FinFET devices for varying L_{gate}

are well controlled by the device structure itself, the fins are left undoped and no HALO implants are necessary. For a fair assessment not the g_{ds} but the intrinsic gain g_m/g_{ds} is considered. The intrinsic gain represents the maximum voltage gain achievable with a single device. Figure 2.5(b) compares FinFET g_m/g_{ds} gate length dependence to planar reference devices at typical analog bias points. Even at short gate lengths the FinFET features improved gain, although its g_m is slightly lower. At longer gate lengths the difference accounts approximately for a factor 3 to 9. Obviously the reason for the high intrinsic gain of FinFETs is the very low output conductance.

The drain current efficiency is defined as ratio of g_m/I_D and limits the achievable bandwidth at given current (power) consumption on circuit level. Figure 2.6(a) shows the efficiency of n-type planar and FinFET device with 90 nm gate length

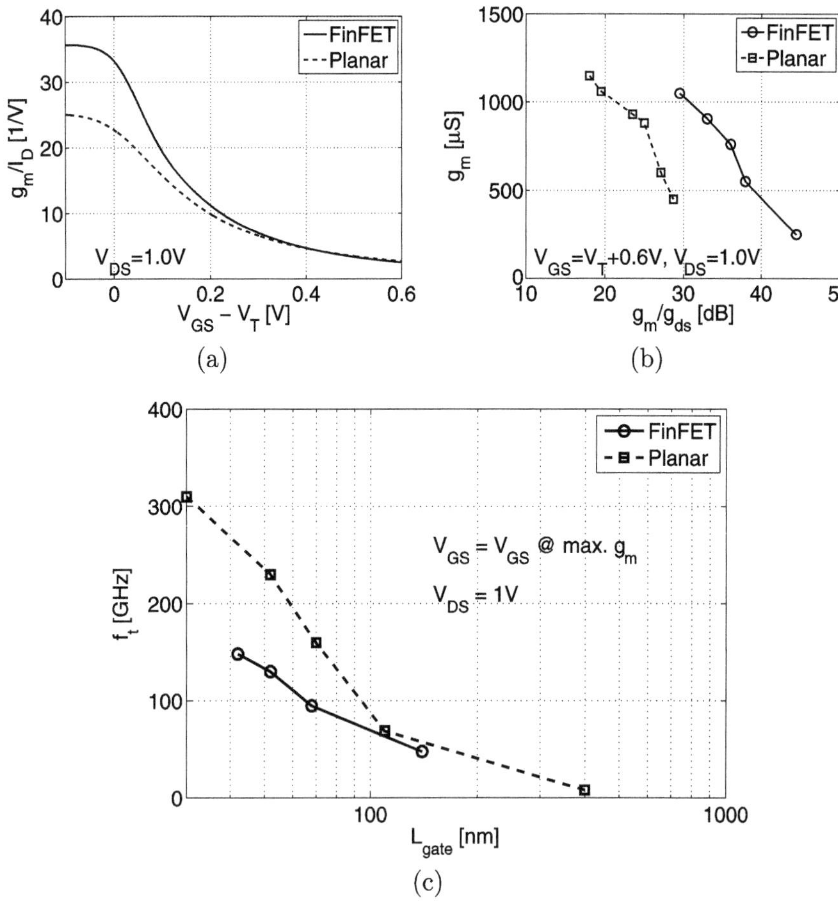

Fig. 2.6 Measured efficiency g_m/I_D of n-type planar and FinFET device with 90 nm gate length (**a**), g_m vs g_m/g_{ds} trade-off (**b**) and transit frequency f_t (**c**) of planar and FinFET devices for varying L_{gate}

versus gate overdrive. Below V_T the FinFET efficiency is significantly improved due to the reduced short channel effects and the corresponding steep sub-V_T slope. At typical low-power analog bias points with moderate overdrive voltage the FinFET is still more efficient, whereas the large series resistances degrade g_m/I_D at high overdrive. The gain versus bandwidth (g_m/g_{ds} versus g_m) trade-off is another important criterion for analog device performance. Figure 2.6(b) compares FinFET and planar devices at high gate overdrive, typically used in high speed applications. Similar to the "low-power" bias points planar devices offer higher maximum g_m, however at much worse gain values. From analog perspective, the high intrinsic gain is a strong argument to use multi-gate devices in future technology nodes, as it overcomes one of the most critical scaling issues in planar bulk CMOS. A quantitative estimation of the resulting benefits for analog circuits is presented later.

For RF circuits the transit frequency f_t and the maximum oscillation frequency f_{max} are key figures-of-merit, defining the unity gain frequencies for current and power respectively. Both quantities relate the achievable transconductance to "parasitics" as gate-source and gate-drain capacitances C_{GS}, C_{GD}. In case of f_{max} also the gate resistance R_G is considered. Figure 2.6(c) shows measured f_t for FinFET and planar reference at maximum g_m. Compared to planar, the f_t of the FinFETs is significantly reduced, mainly due to the higher parasitics: the large source drain resistances R_{SD} limit g_m as shown above and additional fringing capacitances from source drain contact landing pads to the gate contribute to C_{GS} and C_{GD}. The usage of SEG is a limited solution in this case, because any decrease in R_{SD} achieved with wider fins is traded against higher fringing capacitance, see Fig. 2.4. Regarding f_{max} the situation is similar. Nevertheless the currently achievable f_t and f_{max} values well above 100 GHz allow competitive RF circuits in the sub 10 GHz regime as shown later.

The introduction of high-k dielectrics eventually leads to an additional decrease of f_{max} caused by an higher vertical gate resistance. Without careful gate stack engineering the vertical gate resistance can be degraded due to the large number of different interfaces typically present in high-k/metal gate stacks. Figure 2.7(a) illustrates an example containing a SiON interface layer, the main HfSiON high-k dielectric, TiN metal gate electrode, V_T adjust capping layer(s), poly-Si capping and finally the silicide. Of course higher gate resistances affect also noise performance in addition to f_{max}. However this potential issue is not FinFET specific but may occur in any future CMOS technology containing high-k materials.

2.3.2 *Noise Performance*

Thermal and flicker noise are two important noise sources relevant for analog and RF design. The most significant thermal noise source in saturation is the channel. The spectral noise density is given by g_m [45] and consequently independent of device architecture and materials. The increased source drain resistance of the FinFET

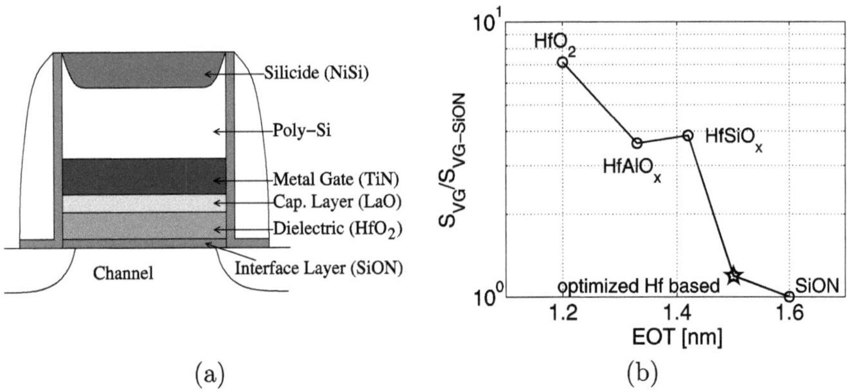

Fig. 2.7 Schematic cross-section of high-k/metal gate stack with exemplary materials (**a**) and relative flicker noise vs EOT trade-off for different dielectric materials (**b**) [65]

may yield slightly higher noise. In contrast to thermal noise, flicker noise strongly depends on technology as explained in Chap. 1. Especially the surface and interface quality is crucial for the flicker noise performance.

Consequently the introduction of high-k dielectrics is a potential source for degraded flicker noise performance due to a high defect density in the high-k dielectric bulk and an increased interface state density [65]. Since flicker noise is very sensitive to many parameters such as thickness and material of interfacial layer and high-k dielectric [46, 47, 66], careful tuning and gate stack engineering is necessary. Exploiting the technology trade-offs combined with lots of process optimization recently a planar bulk CMOS technology featuring high-k/metal gate has been presented [67], achieving competitive flicker noise performance compared to SiON. It is noteworthy that this performance is still worse than what can be achieved with "traditional" SiO_2. Particularly with respect to further scaling of CMOS and the introduction of further novel dielectrics there will be still a material related trade-off between EOT scaling and noise performance as shown in Fig. 2.7(b).

Regarding the FinFET structure there are two potential origins for degraded surface quality and increased flicker noise: the fin sidewalls, which compose about 80% of the channel cross-section, typically feature a (110) crystal orientation which is known to have worse flicker noise behavior [68]. In addition the vertical etch of the fin sidewalls is challenging. Compared to the top surface the roughness of the sidewalls typically is increased. For this reason annealing steps are required to smooth the sidewalls and to achieve sufficient surface quality [31].

Figure 2.8 compares measured spectral noise density of FinFET and planar devices for varying gate overdrive at 1 Hz. The planar reference devices in 32 nm low-power CMOS feature hafnium based high-k dielectric and metal gates [67]. No significant degradation is observed for the FinFET devices with HfSiON dielectric and TiN gate electrode. Both, FinFET and planar are close to ITRS targets for analog, mixed-signal and RF technologies [69].

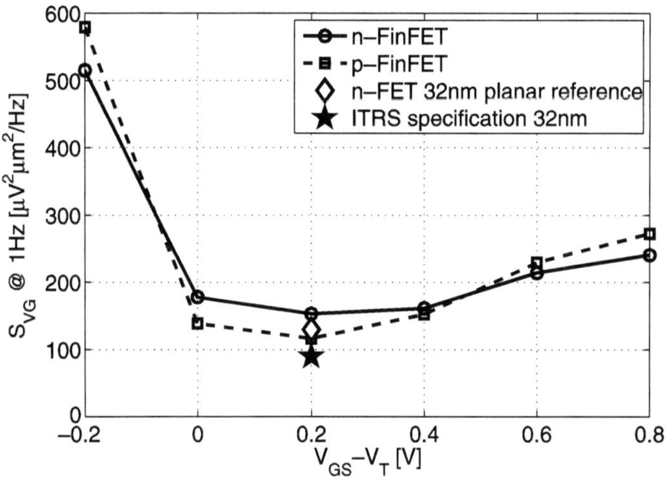

Fig. 2.8 Comparison of measured spectral noise density of FinFET and planar devices for varying gate overdrive at 1 Hz

2.4 Matching Behavior

As shown in Chap. 1 parameter mismatch limits resolution, area and power of analog and mixed-signal circuits. In today's CMOS technologies typically the V_T mismatch induced by random doping fluctuations is dominant, however other effects such as line-edge roughness start to limit scaling of mismatch with EOT according to (1.9).

The introduction of high-k dielectrics impacts matching in a similar way as flicker noise, since both phenomena are related to statistically distributed states and charges. Hence, high-k dielectrics are potential sources for a degradation of matching performance. Pre-existing states and fixed charges at the interface and in the dielectric are randomly distributed and cause local V_T variations comparable random dopant fluctuation. Again, process optimization and dedicated gate stack engineering are required to achieve sufficient matching. On the other side high-k/metal gate offers ways to improve or at least maintain matching performance. The effect of poly depletion (and its variation) is eliminated by the metal gate electrode. Additionally high-k enables scaling of EOT without gate-leakage problems. This EOT scaling enables improved matching constants. Recent 32 nm CMOS technologies prove that high-k is no issue with respect to V_T matching [67].

Compared to planar bulk CMOS the fully depleted FinFET structure offers major advantages: the fin, i.e. the channel region is undoped and the threshold voltage is adjusted by the workfunction of the gate electrode. Since the V_T matching constant shows a square root dependence on the doping level, FinFETs are expected to feature advantageous matching behavior. However, other FinFET specific effects related to fin shape and surface roughness have to be considered with respect to matching [70, 71]. Especially narrow fin devices are affected as illustrated in Fig. 2.9.

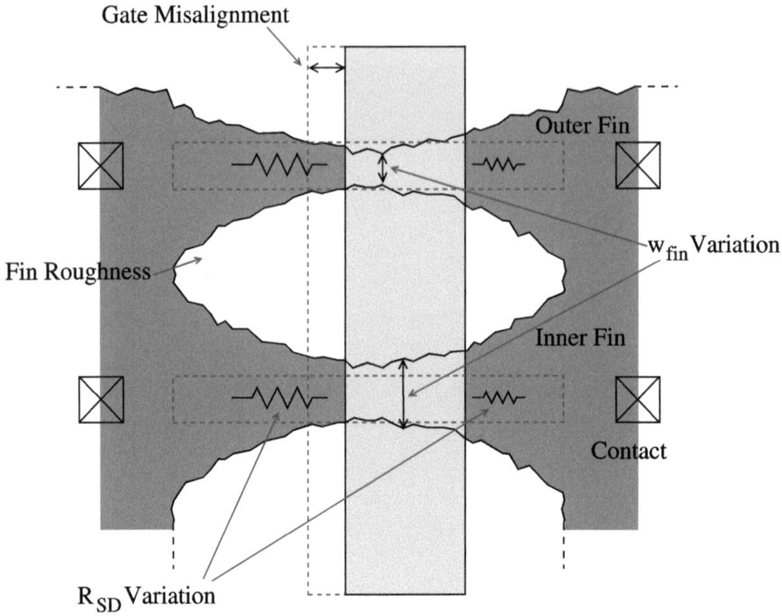

Fig. 2.9 FinFET specific mismatch effects related to fin shape, surface roughness and mask alignment within a multiple fin device

- *Surface roughness*: as explained above, the surface quality of the fin sidewalls is critical. Increased defect and interface state density results in degraded V_T matching.
- *Fin width variations*: the etching of the fins leads to a more or less pronounced broadening of the fin shape in direction to the source and drain landing pads. Particularly in short channel devices this effect has to be considered. Process related, the broadening varies and leads to variations in effective channel width.
- *R_{SD} variations*: (asymmetric) variations in fin width outside of the gate are equal to variations of the source drain series resistances, affecting I_D, g_m etc.
- *Difference between inner and outer fins*: The fin shape of inner fins within a multiple fin device and the two outer fins varies due to the different surrounding: obviously the broadening of the outer fins is different. Lithography requirements enforce different OPC rules for inner and outer fins, which induces additional mismatch.
- *Gate misalignment*: Due to misalignment of masks used in litho process the gate is never perfectly centered between source drain landing pads, resulting in an asymmetric fin shape covered by the gate. Consequently the broadening of the fin is more pronounced on one side of the channel. Since the fin width determines the electrostatics of the device, the asymmetry affects V_T. The amount of V_T shift depends on the direction of misalignment, where the device is more sensitive on the source side.

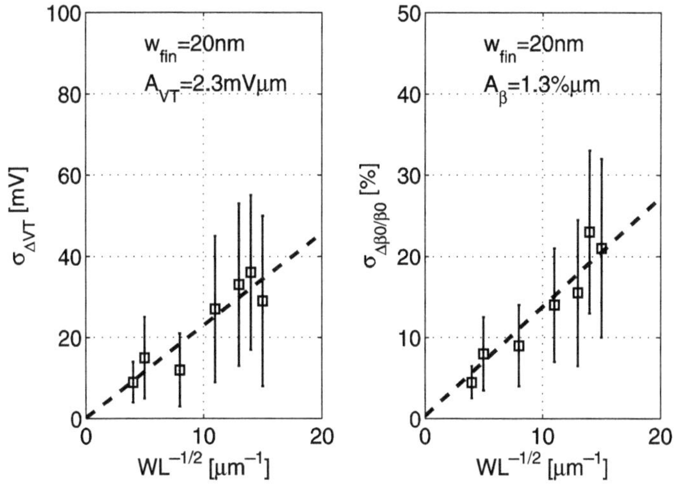

Fig. 2.10 Pelgrom plots of n-type FinFET with 20 nm fin width for V_T and β matching

Table 2.2 Measured matching constants for n- and p-type FinFET with 10 nm and 20 nm fin width

	nFinFET		pFinFET	
	A_{V_T} [mVμm]	A_{β_0} [%μm]	A_{V_T} [mVμm]	A_{β_0} [%μm]
$w_{fin} = 10$ nm	2.7	1.5	2.4	1.1
$w_{fin} = 20$ nm	2.2	1.3	2.3	1.2

Interestingly all effects mentioned above are process related and can be minimized by process optimization in contrast to the dopant fluctuation induced mismatch which is given by statistics. Some of the effects mentioned above depend on transistor layout. Consequently matching optimized layout is required as shown in Chap. 4.

Matching is characterized on identical pairs of n- and p-type FinFETs featuring small fin widths down to 10 nm [54, 72]. To obtain the matching constants, the variances of threshold voltage V_T and current factor β are plotted against device area in the so called Pelgrom plot [42], see Fig. 2.10. The corresponding matching factors for n- and p-type FinFETs are summarized in Table 2.2. Very promising matching behavior is obtained even for narrow fins. Due to the optimization of fin patterning by means of etch process, annealing and OPC no serious degradation of matching behavior at small fins widths is observed as had been reported earlier in [71].

In conjunction with the high intrinsic gain particularly the promising matching behavior is a strong argument for the usage of fully depleted FinFETs in sub 32 nm CMOS technologies. In Chap. 4 a circuit example quantifies the potential performance and/or area benefits enabled by superior matching performance.

2.5 Charge-Trapping

Transient fluctuations of the threshold voltage in the mV regime induced by pronounced charge-trapping is a new phenomenon in emerging CMOS technologies comprising high-k dielectrics, independent of the device structure. To assess the impact of this effect on analog and mixed-signal circuits charge trapping is characterized in detail regarding time constants, stress voltage and temperature dependence. Based on the measurements an equivalent circuit model is derived to enable analog circuit simulation. The measured devices feature high-k/metal gate stack similar to the process published in [54]. The V_T shift is characterized by fast pulsed measurements [73]. After initial V_T measurement the device under test is "stressed" with a gate voltage pulse varying in pulse height and length. To determine amount and recovery of the V_T shift the threshold voltage is measured after the pulse with variable delay.

The approach to model transient V_T shifts derived in this work is shown in Fig. 2.11. The ideas presented in [48] and [50] are extended to cover asymmetric trapping and de-trapping time constants. The V_T shift is represented by a voltage controlled voltage source V_{shift} in series with the gate. The controlling voltage itself is derived by an equivalent sub-circuit model: a voltage controlled voltage source V_{ss} applies the steady state V_T shift to an RC network. Steady state is defined as the V_T shift which is measured at sufficient large (>1 ms) pulse width where no more dependence on pulse width is observed, i.e. all traps are filled. The RC network models the distributed time constants resulting from the distributed traps, comparable to

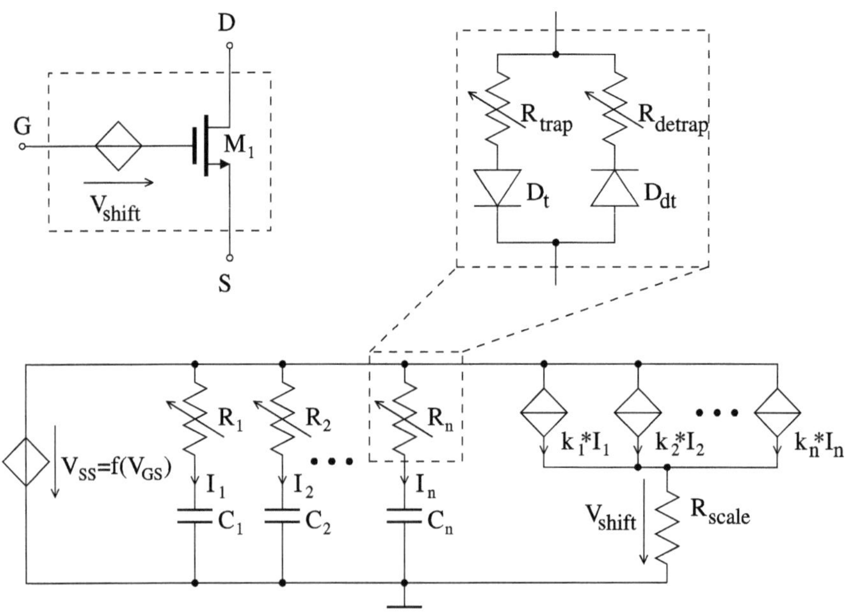

Fig. 2.11 Modeling of V_T shift as voltage controlled voltage source and implementation details of the equivalent circuit for the control voltage

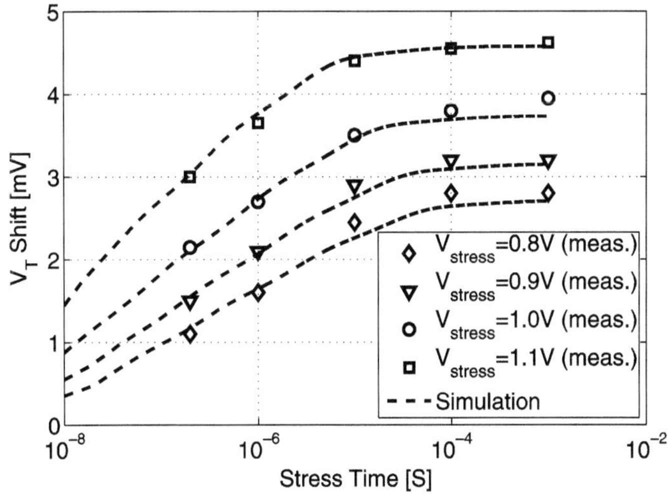

Fig. 2.12 Measured and simulated V_T shift for varying gate pulse height and length

simple flicker noise models. In contrast to [48] the resistors are voltage controlled to represent voltage dependent filling of traps. Different resistors are used for charging (trapping) and discharging (de-trapping). This approach allows to model asymmetric time constants for field enhanced trapping (positive gate pulse) and de-trapping (zero gate pulse or negative gate pulse). The RC current contributions $I_1, \ldots, I_n$ are weighted with scaling factors k_i in controlled current sources. The sum of the weighted currents induces a voltage drop over a the resistor R_{scale} which yields the final shift voltage. R_{scale} is a variable model parameter which can be used to set up different hysteresis scenarios.

To enable circuit simulation a set of model parameters is extracted from measurements. The V_{GS} dependence of the steady state V_T shift is modeled by means of a 4th order polynomial source V_{ss}. To reflect the hysteresis timing behavior properly 6 weighted[2] RC elements are used, covering trapping time constants from 10^{-8} s to 10^{-3} s and de-trapping time constants from 10^{-6} s to 10^{-1} s, respectively.

Measurement and simulation results are shown in Fig. 2.12 for varying gate pulse height and length. Three main statements are derived from these results: very fast traps with time constants below the µs regime have to be considered. Hence, the time constants of the effect are in a relevant range for typical analog and mixed-signal applications. Furthermore a strong dependence on the pulse height is observed, as shown in Fig. 2.13. As a consequence the mismatch of devices with different operation history will be significant. Finally the model enables a prediction of hysteresis behavior in circuit simulation with sufficient accuracy. The recovery behavior of the V_T shift is depicted in Fig. 2.14. As expected the de-trapping under zero gate bias occurs at much higher time constants than the field enhanced trapping. The differ-

[2] $k_1 = 1$, $k_2 = 1$, $k_3 = 1$, $k_4 = 0.84$, $k_5 = 0.15$, $k_6 = 0.05$.

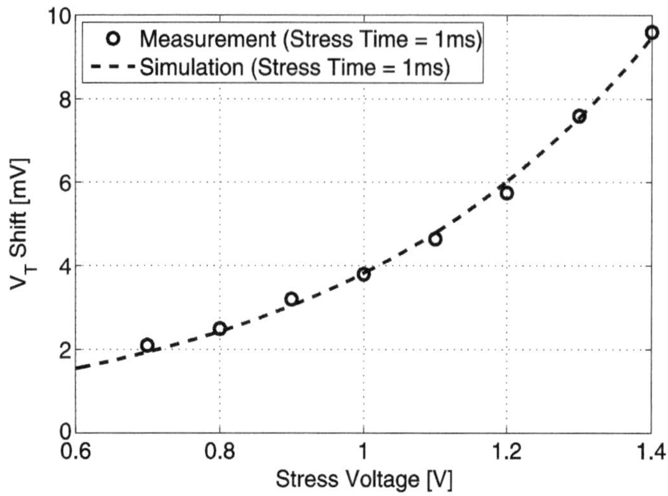

Fig. 2.13 Measured and simulated steady state V_T shift for varying gate pulse height

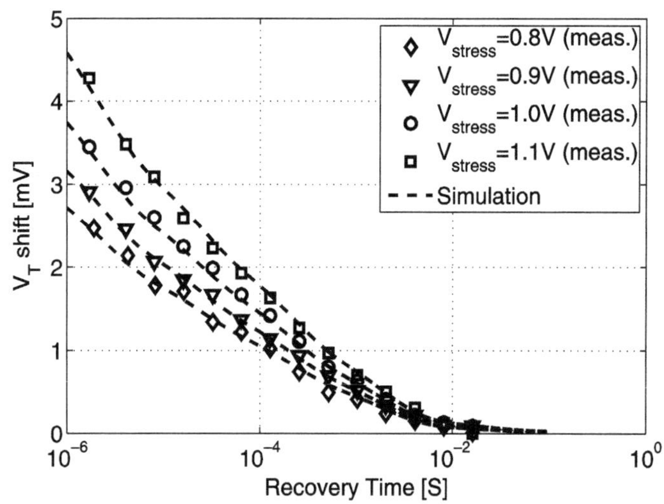

Fig. 2.14 Measured and simulated recovery of V_T shift for varying gate pulse height and measurement delay

ence amounts to about three orders of magnitude. It is noteworthy that negative gate pulses can accelerate de-trapping.

Measurement results reveal minor temperature dependence, no significant thermal activation of traps is observed. Therefore temperature effects are neglected in the circuit model. Overall, measurements and simulations are in close agreement. The model derived here serves as basis for the assessment on circuit level, using the scalability of the maximum V_T shift to simulate different hysteresis scenarios. Although the absolute values of V_T hysteresis ($\approx$5 mV at 1 V stress) are representative for state-of-the-art high-k dielectrics, further process optimization may yield even lower levels [74]. Nevertheless the impact of V_T hysteresis on analog and mixed-signal circuits is studied in detail in this work. The motivation is twofold: one main goal of the investigation is to derive specifications for maximum tolerable V_T shift from analog perspective intended as feedback for technology development. On the other hand an early assessment of the impact on circuit level and development of corresponding countermeasures is required with regard to further scaling of CMOS technologies comprising novel material systems with possibly even higher V_T instabilities.

2.6 Self-Heating

The power dissipation in transistors leads to a local temperature increase, called self heating. Temperature changes affect transistor parameters such as carrier mobility or threshold voltage. At high gate-source and drain-source bias, i.e. at high current densities the reduction of the mobility is the dominant effect, resulting in a reduction of the drain current. As mentioned earlier the effect is particularly severe in SOI technologies due to the low thermal conductivity of the surrounding oxide (mainly the buried oxide layer) that acts as thermal insulator.

The impact of self heating on the output characteristics of an n-type FinFET is illustrated in Fig. 2.15(a). Pulsed measurements at different temperatures are compared to static measurement at room temperature. The pulse width is chosen very small in order to prevent self heating in these measurements. The DC measurement in contrast suffers from self heating and represents a kind of mean value of the pulsed measurements: at low V_{DS} values, i.e. at low power density it follows the room temperature curve, whereas at high V_{DS} values i.e. at high power density it converges to the 75°C curve. The maximum DC current is reduced by nearly 10%,

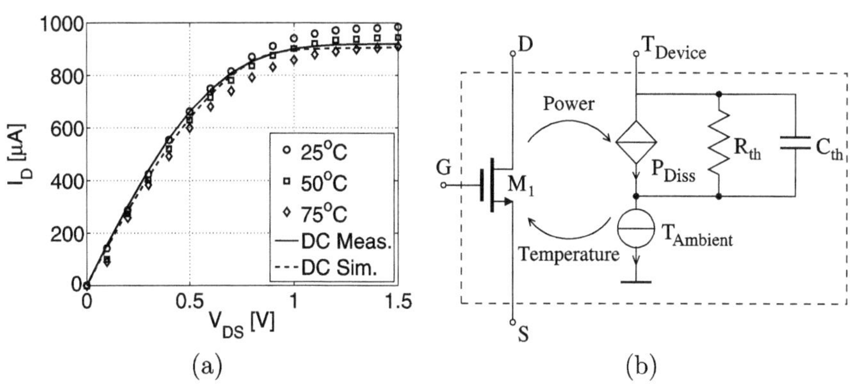

Fig. 2.15 Comparison of pulsed and DC I_D–V_{DS} measurements/simulation (**a**) and equivalent circuit model for self heating (**b**)

equivalent to a temperature rise of about 50°C. However, the power density in this example is larger than in typical analog use cases.

For analog circuit simulation self heating is modeled by means of electro-thermal coupling in a simplified equivalent circuit [75], as shown for a n-type FinFET in Fig. 2.15(b). The dissipated transistor power is sensed and applied via a controlled source to a thermal RC network comprising a thermal resistance R_{th} and a thermal capacitance C_{th}. R_{th} is defined as temperature increase per dissipated power: $R_{th} = \Delta T/P$. C_{th} is used to fit the time dependence. The model can be easily extended to cover multiple time constants [76]. The thermal parameters R_{th} and C_{th} are extracted by pulsed measurements, RF small signal measurements [77] or device simulations [76]. For typical FinFET device dimensions R_{th} is in the range of 10°K/mW to 100°K/mW, the respective time constants are in the 10–100 ns regime [50, 76]. Figure 2.15(a) shows that the simulations fit well the measured data.

Similar to charge trapping self-heating yields history dependent transistor behavior and potential dynamic mismatch effects. The impact on analog and mixed-signal circuits is discussed in Chap. 4. Since self-heating is strongly affected by the device geometry the scaling behavior of self-heating is analyzed next. To avoid the need for complex and time consuming device simulations a simplified approach is used [78]. The complex 3D FinFET structure is partitioned in individual basic blocks in such a way that the thermal resistance of the basic blocks can be analytically calculated. The model includes gate stack, source and drain landing pads, source, drain and gate contacts as main heat conductors and is calibrated with device simulations and measurements. In this way the impact of geometry variations can be easily assessed.

The fin width is a key scaling parameter for FinFETs due to its impact on electrostatic integrity. The dependence of the thermal resistance on the fin width for constant gate length of 45 nm and 8 fins in parallel is shown in Fig. 2.16(a). As expected, with decreasing fin width, i.e. with decreasing silicon volume the thermal resistance increases up to a factor of two. However, in a realistic scaling scenario not only the fin width but almost all dimensions are reduced. In Fig. 2.16(b) the thermal

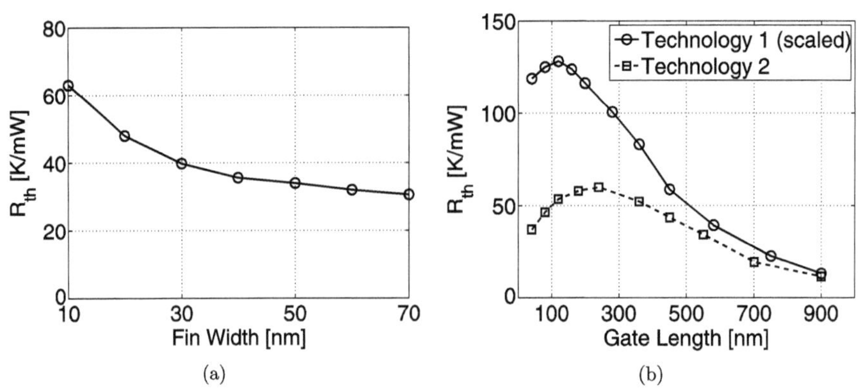

Fig. 2.16 Simulated scaling behavior of thermal resistance with regard to fin width (**a**) and all relevant dimensions (**b**)

resistance of two FinFET technologies is compared. Technology 2 refers to typical dimensions as used in [6]. Technology 1 represents a down-scaled technology, assuming a scaling factor of 0.7 applied to minimum gate length, fin width, fin pitch, contact and metal dimensions, contact density and thickness of buried oxide layer. The thermal resistance increases up to a factor of 2 in this example. The conclusion drawn from these simulations is that the impact of self-heating on device behavior is increasing with technology scaling and has to be considered in design and layout.

Chapter 3
High-k Related Design Issues

High-k gate dielectrics will be a key component of future planar or multi-gate CMOS technologies. The potential impact on analog device properties is summarized in Chap. 2. This chapter deals with high-k related circuit design issues from analog and mixed-signal perspective: increasing flicker noise and charge trapping induced dynamic V_T variations are among the main concerns related to the introduction of high-k dielectrics [47].

Although state-of-the-art high-k dielectrics achieve moderate flicker noise and hysteresis levels it is important to understand the fundamental impact on circuit design: future CMOS technologies may require second generation high-k dielectrics based on novel material systems to enable further EOT scaling. These gate dielectrics will feature even higher k values but eventually worse electrical properties [65].

3.1 Flicker Noise

Since flicker noise is a well known phenomenon in CMOS analog circuit design [61] the discussion here is restricted to some very fundamental considerations. In general, the impact of flicker noise depends on the actual application.

3.1.1 Linear Analog Circuits and Converters

In linear analog circuits and A/D & D/A converters the rms value of the noise is the quantity of interest which determines the achievable signal-to-noise ratio (SNR) or the dynamic range [62]. The rms value is given by the integral of the spectral noise power density over the system bandwidth. Consequently the contribution of flicker noise is determined by the relation of system bandwidth to flicker noise corner frequency. Assuming a system bandwidth BW starting at 1 Hz, a spectral flicker noise

M. Fulde, *Variation Aware Analog and Mixed-Signal Circuit Design in Emerging Multi-Gate CMOS Technologies,* Springer Series in Advanced Microelectronics 28,
DOI 10.1007/978-90-481-3280-5_3,

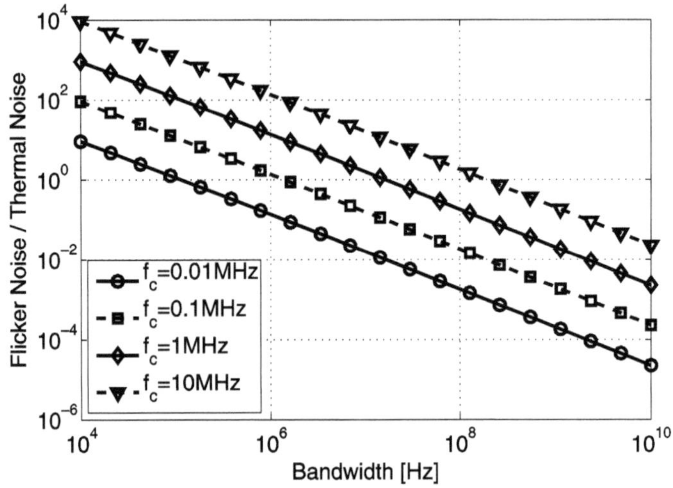

Fig. 3.1 Calculated relation of flicker to thermal noise contribution versus bandwidth for different flicker noise corner frequencies

density of a/f and a spectral thermal noise density of b, the rms noise value $\overline{v_n^2}$ is given by

$$\overline{v_n^2} = \int_1^{BW} \frac{a}{f} df + \int_1^{BW} b df. \tag{3.1}$$

With the flicker noise corner frequency f_c the relation of flicker noise contribution $\overline{v_{1/f}^2}$ to thermal noise contribution $\overline{v_{th}^2}$ can be calculated as

$$\frac{\overline{v_{1/f}^2}}{\overline{v_{th}^2}} = f_c \frac{\ln(BW/1\text{ Hz})}{BW}. \tag{3.2}$$

The relation is illustrated in Fig. 3.1. Especially in systems with low or moderate bandwidth, e.g. voice coding or GSM mobile phone applications a potential increase of flicker noise significantly degrades the signal-to-noise ratio. To maintain the required dynamic range, i.e. to reduce the impact of flicker noise, the device area has to be increased according to (1.12). Increased device area obviously results in increased circuit area and power consumption (or degraded speed). Typically degraded flicker noise performance translates linearly into area and power overhead [45, 66].

3.1.2 Voltage Controlled Oscillator

Voltage controlled oscillators (VCOs) are important building blocks of nearly all communication systems [79], used e.g. in phase-locked-loop circuits for clock generation. Besides center frequency, power consumption and tuning range the phase noise is a key figure-of-merit of integrated VCOs [80]. Roughly speaking, phase noise describes the spectral purity of the carrier.

Although the oscillating frequency typically exceeds the flicker noise corner frequency by several orders of magnitude, flicker noise is an important device parameter for the design of VCOs: non-linearities and asymmetries cause an up-conversion of low-frequency noise, degrading the phase noise close to the carrier frequency [81]. Far from carrier, i.e. at large offset frequencies up-converted thermal noise dominates. In contrast to linear circuits the impact of flicker noise on VCO phase noise can be reduced in two ways: a lowering of flicker noise at device level again results in power and eventually area overhead. In integrated CMOS LC oscillators typically the inductor limits the area, whereas in ring oscillators the active devices determine the circuit area. Another option for the reduction of close to carrier phase noise is a dedicated design optimization to suppress the up-conversion mechanisms of low-frequency noise [81]. Ensuring symmetric oscillator waveforms is one example for reduction of up-conversion mechanisms by design measures.

3.1.3 Flicker Noise Reduction Techniques

Lowering flicker noise by enlarging the active transistor area typically results in significant area and power overhead. Several design techniques have been proposed to minimize the impact of flicker noise on circuit level and to reduce flicker noise itself on device level.

3.1.3.1 Chopper-Stabilization and Auto-Zeroing

Chopper-stabilization and auto-zeroing are circuit techniques to reduce the impact of flicker noise on circuit level. The principle of chopping is shown in Fig. 3.2.

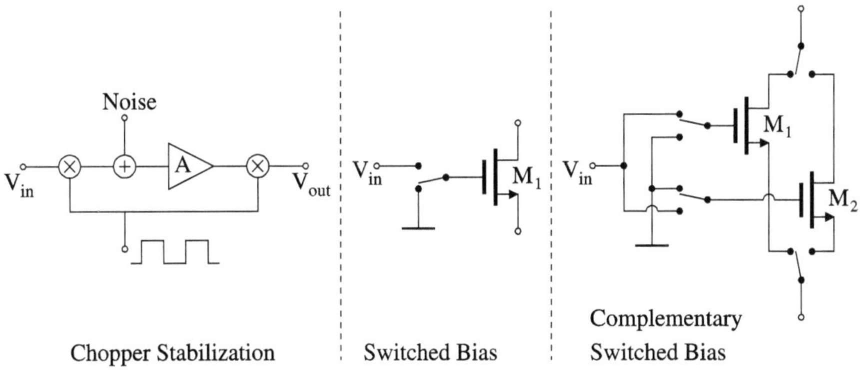

Fig. 3.2 Different design techniques to reduce the impact of flicker noise

Chopping is based on up-conversion of low-frequency noise [82]. The input signal is multiplied by a square-wave signal with the frequency f_{chop} leading to a translation (modulation) of the signal to the odd harmonics of the chopping signal. The input referred amplifier noise is modelled by an equivalent noise source which is added to the signal spectrum in front of the amplifier. The up-converted signal spectrum is not distorted by the low frequency noise components. A second multiplier is used to demodulate the wanted signal after amplification, whereas the noise is up-converted. If the chopping frequency is higher than the signal bandwidth and the flicker noise corner frequency the inband noise is significantly reduced. Disadvantages of chopping are increased thermal noise due to the switching and increased bandwidth requirements on the amplifier [62].

In switched capacitor circuits offset voltages are compensated using the so called auto-zeroing technique. The offset is stored on a capacitor in the first phase and then subtracted in the second phase, see Sect. 3.2.2.2 and Fig. 3.11. Auto-zeroing reduces also low frequency noise components, which can be interpreted as time dependent offset voltage [83]. The efficiency of flicker noise reduction increases with the ratio of clock frequency to flicker noise corner frequency.

3.1.3.2 Switched Bias

There are also methods to reduce the intrinsic transistor flicker noise. Flicker noise is related to trapping and de-trapping of minority carriers in oxide or interface traps. Switching a MOSFET periodically from inversion to accumulation state modulates the trapping and de-trapping time constants and lowers the probability of carrier fluctuation, thus noise is reduced [84]. The idea is shown in Fig. 3.2. It is noteworthy that switching a FET on and off with 50% duty cycle yields a 50% noise reduction per definition, since a device in off-state contributes no noise. However, the modulation of trapping and de-trapping time constants reduces the noise also in the on-state. The implementation of this technique is straightforward in circuits where bias current is needed only during a certain time interval, e.g. in oscillators [85]. An extended switched bias technique can be applied in linear circuits where constant bias currents are required [86]. In the so called complementary switched bias scheme a second transistor is required. The input voltage is periodically switched between the two devices, the other one is connected to ground (in case of an nFET), see Fig. 3.2. Although a remarkable reduction of flicker noise is achievable, the concept suffers from increased thermal noise due to the switches in the signal path and from switching noise [87]. Recently an improved switched bias technique was proposed, making use of forward substrate biasing during off-state [88]. The noise reduction is improved by more than one order of magnitude. However, the forward biasing of single devices results in significant increased power consumption [89]. Moreover, if applied to nFETs a triple well option is required to separate the substrate potentials. In fully depleted SOI devices obviously no local substrate bias can applied.

Summarizing the considerations on flicker noise some important statements are concluded. Depending on the application flicker noise can dominate the overall noise performance. Compensating increased flicker noise levels either by enlarging device area or by dedicated circuit techniques always implicates a power or area penalty. In latter case reduced flicker noise is traded against higher thermal noise and switching noise. Thus, keeping flicker noise at moderate levels is a strong prerequisite for the introduction of high-k dielectrics.

3.2 Transient V_T Variations and Hysteresis Effects

In this section the impact of charge trapping induced dynamic V_T variations on analog and mixed-signal circuits is discussed. As mentioned in Chap. 1 pronounced charge trapping or hysteresis is a new effect related to the introduction of high-k dielectrics. Although the hysteresis levels reported for state-of-the-art high-k dielectrics are in the range of few mV the effect is investigated in detail here. Even small dynamic V_T variations can degrade the performance of high resolution analog and mixed-signal circuits in case of small supply voltages [50]. The main target of this discussion is to provide a basic assessment on the impact of dynamic V_T variations in analog and mixed-signal circuits. The scalable circuit model presented in Chap. 2 is used to derive specifications for tolerable hysteresis levels. Finally countermeasures on circuit and block level are proposed.

3.2.1 Linear and Continuous Time Building Blocks

In general, all devices with non-zero gate bias are affected by charge trapping. However, from DC perspective charge trapping acts like a small constant V_T shift. A deterministic, constant V_T shift can be easily considered in the design by adjusting bias voltages and currents or W/L ratios. Thus, charge trapping and hysteresis is no concern for all kinds of circuits with a constant operating point, e.g. biasing circuits or current mirrors. Circuit behavior is only affected if devices are exposed to different operating history in terms of transient gate signals, e.g. in differential amplifiers or if mismatch of hysteresis parameters needs to be considered.

3.2.1.1 Single Ended Operational Amplifier

Assuming reasonable open loop gain the input signals of an operational amplifier have to be very small to prevent clipping of the output stage, regardless of open or closed loop operation. In any case the differential OpAmp input signal has to be smaller than the ratio of supply voltage to open loop gain ($V_{in} < V_{DD}/A_0$), typically less than few mV.

Hence, the resulting transient mismatch of the differential input pair is very small: the V_T shift depends exponentially on the stress voltage and the maximum steady state V_T shift (measured at $V_{stress} = V_{DD}$) is typically in the order of several mV as shown in Chap. 2. Assuming a differential input signal of few mV the transient mismatch of the input pair is in the range of few μV. The resulting impact on the main OpAmp figures-of-merit like open loop gain, common-mode and power supply rejection ratio, gain-bandwidth product, slew rate and settling behavior is negligible for all practical purposes: according to simulations with a two stage Miller compensated OpAmp as shown in Fig. 4.6 the quantities listed above are affected by less than 0.1% for a maximum steady state V_T shift below 30 mV.

The worst case scenario regarding the differential OpAmp input is a full scale pulse wave signal in closed loop configuration as shown in Fig. 3.3. During OpAmp settling there is as significant differential "glitch" signal at the negative input node V_{inn}, see Fig. 3.4(a). Due to the asymmetric time constants for trapping and detrapping a charge pump like mechanism causes a rising V_T shift at the negative OpAmp input, equivalent to an offset voltage. The distributed time constants yield a $\log(t/\tau)$ time dependence as shown in Fig. 3.4(b). Compared to the maximum steady state value, the offset saturates at much lower values, due to the low duty cycle of the "stress" pulse at the gate. The actual impact on the output waveform (in time and frequency domain) therefore is almost negligible. In case of a sinusoidal input waveform there is no settling of the OpAmp, as long as the signal frequency is smaller

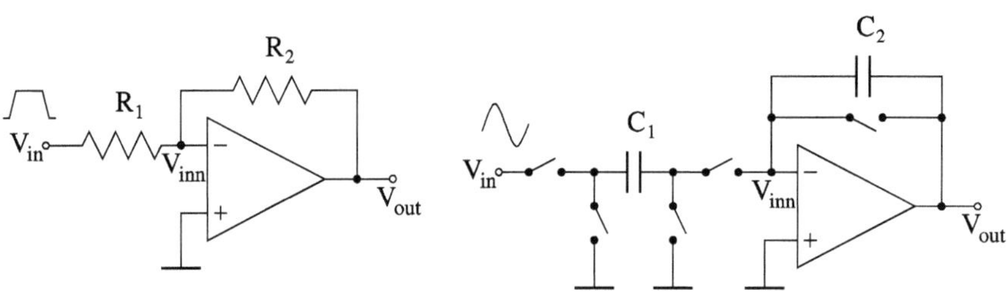

Fig. 3.3 Schematic of single ended, continuous time (*left*) and switched capacitor (*right*) closed loop amplifier configuration

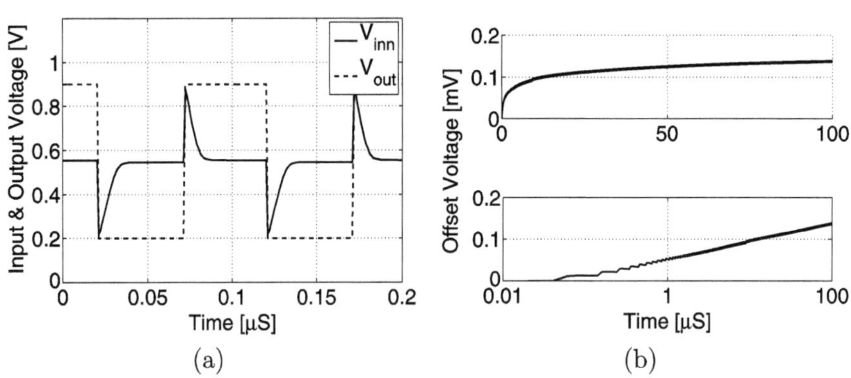

Fig. 3.4 OpAmp output and differential input voltage in feedback configuration with pulse wave input signal (**a**) and corresponding transient offset voltage (**b**)

than the OpAmp bandwidth. Consequently there is no significant modulation of the OpAmp input node V_{inn} and the output waveform is even less affected. It is noteworthy that also the device at the positive OpAmp input shows a tiny transient offset which is caused by the coupling from the differential pair tail node.

3.2.1.2 Fully Differential Operational Amplifier

In fully differential OpAmp configurations both input devices are exposed to antisymmetrical input pulses, assuming real fully symmetrical input signals. Consequently, both transistors will show the same transient V_T shift, assuming perfect matching. In this case the transient V_T variation causes no dynamic offset voltage but a common mode signal. Assuming reasonable common mode rejection this signal is attenuated by several tens of dB. In other words fully differential amplifiers are even less affected by hysteresis effects. However, any mismatch in the charge trapping behavior of the input devices results in transient offset. Mismatch of hysteresis parameters is not considered here due to missing measurement data. Since hysteresis is attributed to traps and defects which are randomly distributed, an area dependent variation of hysteresis parameter should be assumed, similar to flicker noise [90]. This aspect needs to be addressed in further work.

3.2.2 Non-Linear and Discrete Time Building Blocks

As shown above pulsed waveforms yield transient V_T shift and offset voltage in continuous time OpAmp configurations. In non-linear open loop or discrete time systems like comparators or switched capacitor amplifiers this effect occurs even more pronounced.

3.2.2.1 Switched Capacitor Amplifier

Figure 3.3 shows a switched capacitor implementation of a feedback amplifier. The signal V_{inn} at the negative OpAmp input is comparable to a continuous time implementation, regardless of the actual waveform of V_{in}. However, duty cycle and amplitude of the stress pulse is larger in this case. Depending on the clock frequency the resulting dynamic offset voltage is not negligible any more. Although a dynamic offset voltage in the mV range is almost not visible in the time domain, in the output spectrum a significant increase of noise floor at low frequencies is observed. The frequency dependence is comparable to flicker noise. The simulated output spectrum for a 2 MHz sine wave input and a clock frequency of 20 MHz is shown in Fig. 3.5. The signal amplitude as well as the harmonics are not affected by the dynamic V_T shift, however the signal-to-noise ratio is degraded in this example by about 1 dB.

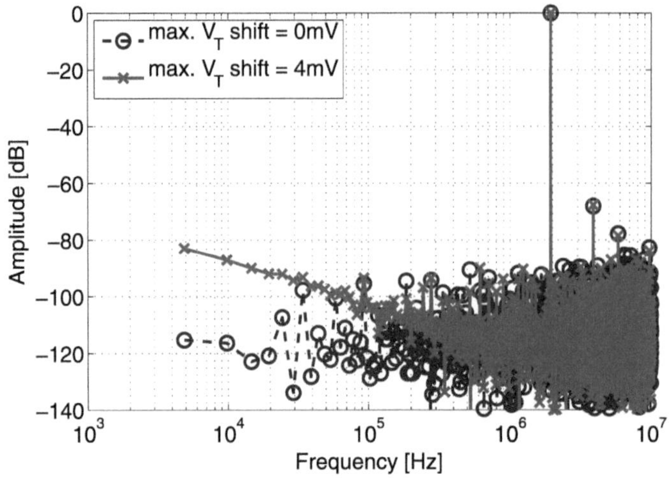

Fig. 3.5 Simulated output spectrum of switched capacitor amplifier with 2 MHz sine wave input and a clock frequency of 20 MHz with and without hysteresis effect

In fully differential switched capacitor amplifiers the charge trapping effect again causes no transient offset but a small common mode signal which is suppressed by several tens of dB. A simulation under same conditions as above yields no observable degradation of SNR. Again, mismatch of hysteresis parameters is neglected.

3.2.2.2 Comparator

Comparators are high gain amplifiers in open loop configuration. Linearity is no concern, since comparators are intended to deliver a digital output. Typically several pre-amplifying gain stages are used to generate a sufficient signal swing for the final latch, see Fig. 3.6. Regarding transient V_T variations comparators are operated in the worst case scenario: the input signal is only limited by the supply or reference voltage, i.e. the input devices are not stressed by small glitches but large voltage pulses. Depending on the history of input voltages the transient V_T mismatch can even reach the maximum steady state value.

Incorrect Comparator Decisions The dynamic V_T mismatch can cause incorrect comparator decisions, if the differential input voltage $V_{in} = V_{inp} - V_{inn}$ is rapidly changed from a large value to a small value with equal sign. The mismatch created under large differential input voltage partly remains in the following decision. Figure 3.7 illustrates this effect. Here a three stage comparator as shown in Fig. 3.6 is simulated. The final latch is not considered. The differential input voltage is switched from +800 mV to +3 mV. In case of zero V_T shift the comparator is able to follow the input signal and gives the correct decision, see Fig. 3.7(a). However, a wrong output is observed using the correct (i.e. the measured) V_T shift values in the circuit model.

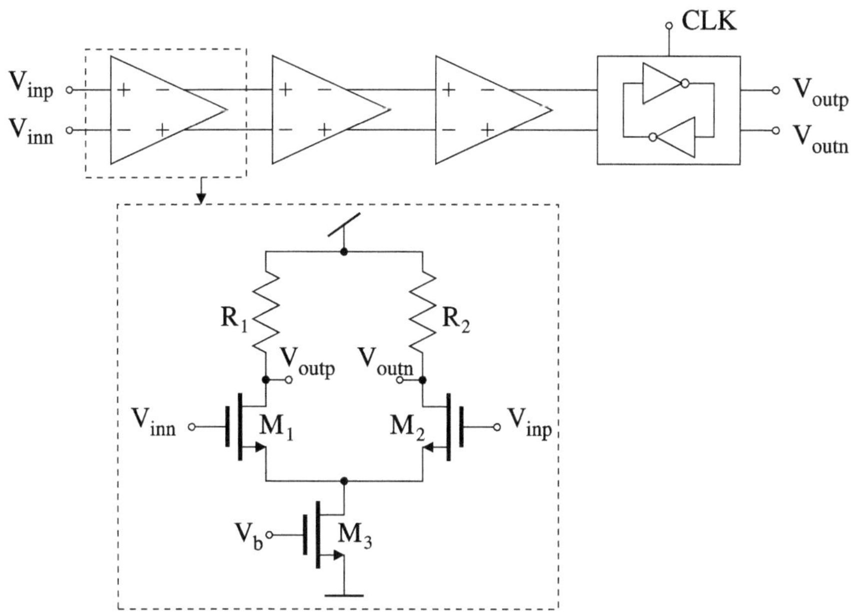

Fig. 3.6 Schematic of three stage comparator with final latch and schematic of single gain stage

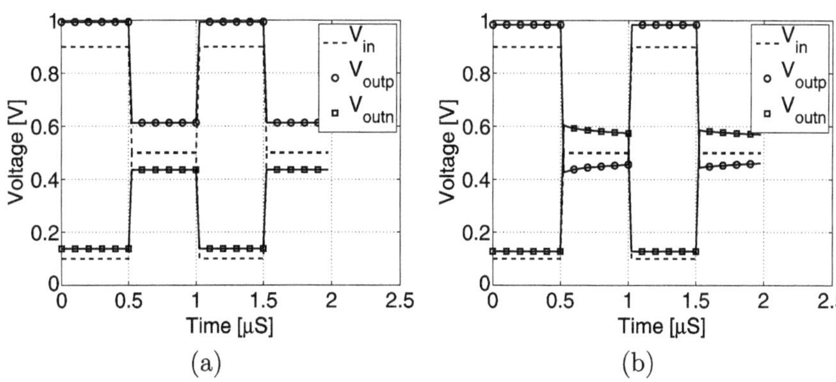

Fig. 3.7 Transient comparator response with maximum V_T shift = 0 mV (**a**) and with maximum V_T shift = 4 mV (**b**)

An analytical error prediction is not possible, since the actual V_T mismatch depends on the operation history, not only on the previous input value. If the maximum steady state V_T shift is assumed as worst case, the actual input value has to be larger than this value to guarantee a correct decision. Hence the comparator resolution is limited to the maximum steady state V_T shift.

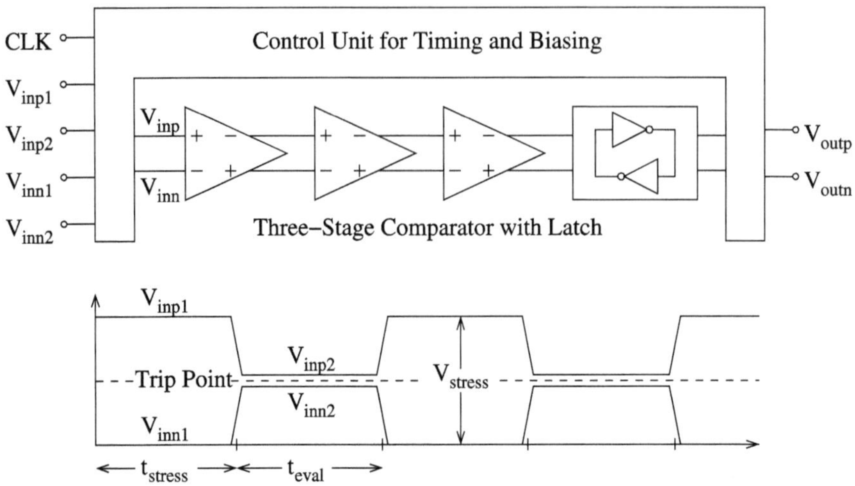

Fig. 3.8 Schematic of comparator testbench and corresponding timing diagram

Comparator Testbench To verify this effect on silicon an integrated comparator testbench is realized in FinFET technology as shown in Fig. 3.8. Internal multiplexers are used to generate accurate input voltage pulses from different external precision voltage supplies. The timing unit allows to adjust the stress and evaluation time t_{stress} and t_{eval}. The comparator itself consists of three pre-amplifier stages and differential latch. The output is latched and measured at the end of evaluation time. Since the comparator is distorted by noise, the digital output is averaged over a large number of clock cycles to calculate the bit error rate. $V_{in2} = V_{inp2} - V_{inn2}$ is calibrated as minimum input voltage that is detected by the comparator with a certain minimum bit error rate. This minimum bit error rate is limited by the noise level. The stress voltage $V_{in1} = V_{inp1} - V_{inn1}$ is varied. The measured bit error rates for varying stress voltage and different stress times are shown in Fig. 3.9. All measurements are performed with a supply voltage of 1.0 V and a common mode level of 0.6 V. V_{in2} is set to 2 mV, corresponding to a bit error rate below 0.5%. For differential stress voltages between 0.2 V and 0.6 V no change in bit error rate is observed. With further increasing stress voltage the bit error rate rises exponentially due to the increasing transient mismatch. The bit error rate is doubled at a stress voltage of 1 V which corresponds to nominal V_{DD}. At 2 V differential input the error probability is almost one. The bit error rate is comparable at 100 μs and 1000 μs stress time, however it is reduced at 1 μs, which indicates that not all traps are filled within this pulse width.

Figure 3.10 shows the bit error rate at constant stress voltage as function of stress time with constant eval time = 1 μs (a) and as function of eval time with constant stress time = 1000 μs (b). Obviously the main part of traps is filled within 10 μs, since longer stress time yields no significant further degradation of bit error rate. The contribution of traps with larger time constants seems negligible. Increasing the evaluation time improves the bit error rate due to the detrapping which lowers the

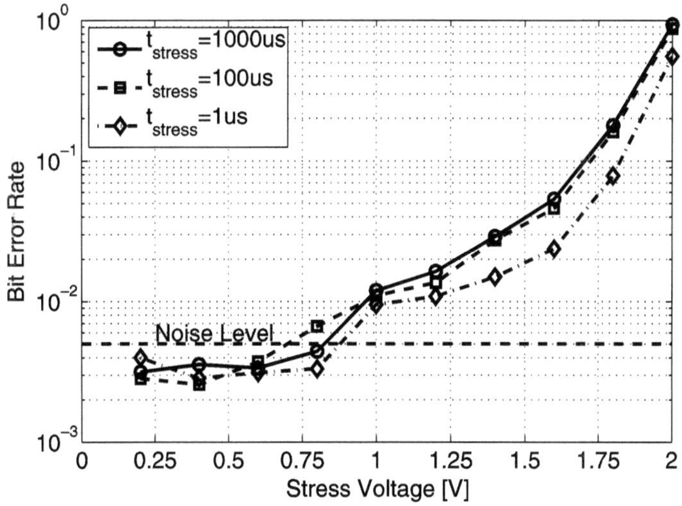

Fig. 3.9 Bit error rate as function of stress voltage for varying stress time

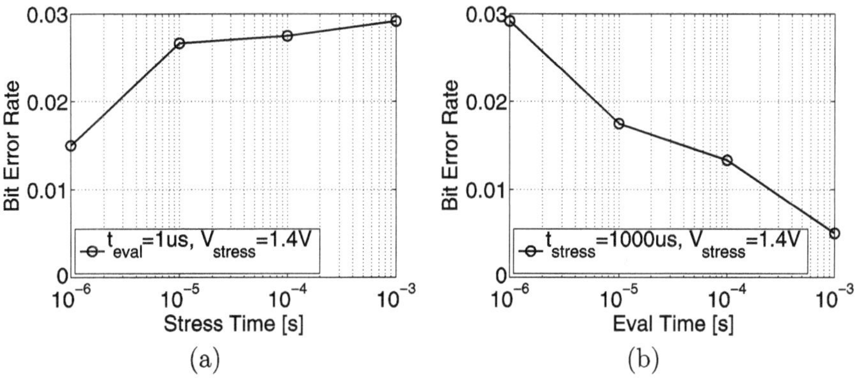

Fig. 3.10 Bit error rate as function of stress time with constant eval time = 1 μs (**a**) and as function of eval time with constant stress time = 1000 μs (**b**), stress voltage = 1.4 V

transient mismatch. At 1 ms evaluation time the transient mismatch is almost fully recovered and the bit error rate is again limited by noise.

The experimental results prove that charge trapping induced transient V_T shifts can cause incorrect comparator decisions. The results are in line with single device measurements and simulations. A significant increase in bit error rate is observed, especially at large input values close to V_{DD}. Different approaches to compensate for the effect are discussed next.

Offset Compensation & Auto Zeroing The so called auto zeroing is a well known technique to compensate static mismatch or offset in switched capacitor circuits as mentioned above. The principle of operation is illustrated in Fig. 3.11.

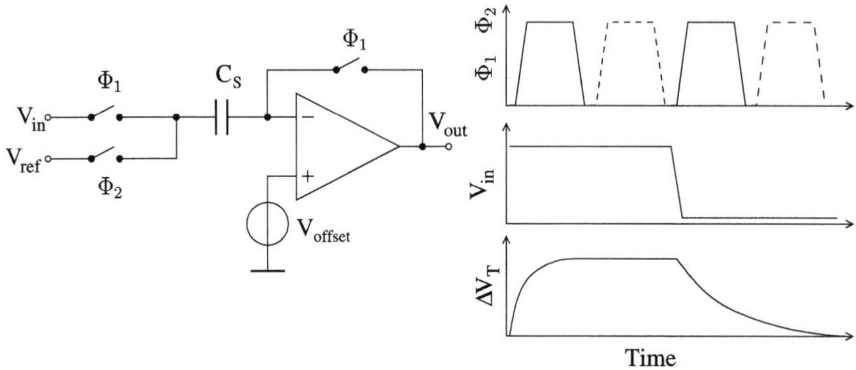

Fig. 3.11 Schematic of offset compensated comparator with timing diagram

During the sampling phase Φ_1 the difference of input and offset voltage is stored on the sampling capacitor. In the evaluation phase Φ_2 the offset voltage is added again to the negative input and thus cancelled out. Auto zeroing or correlated double sampling is also known to reduce the impact of low frequency noise [83] which can be regarded as time dependent offset. In a similar way transient mismatch is reduced. However, the accuracy of auto zeroing is limited by the ratio of the hysteresis time constants to the clock period. The sampling clock has to be much faster than the trapping or de-trapping time constants, otherwise the offset may be under or overcompensated. Since the transient mismatch depends on the operation history the offset compensation process is necessary for every comparator decision, which is not feasible in some applications.

Switched Comparator Input An alternative approach is presented next. The concept is illustrated in Fig. 3.12. Each transistor of the differential input pair in the gain stage of the comparator is split up into two equally sized devices. One device is always connected to the positive (negative) input, whereas the other one is switched between positive and negative (negative and positive) input after each decision. Thus, the positive input device consists of M_{1a} and M_{1b} for one clock cycle and of M_{1a} and M_{2b} for the next clock cycle. Consequently the mean dynamic V_T shift of positive and negative input device is equal and the dynamic offset is canceled out, assuming that the sampling time is smaller than the (de)trapping time constants.

Figure 3.13 shows the simulated comparator response with switched input under worst case conditions (1st input value $= +V_{DD}$, 2nd input value $= +3$ mV). Two versions of the comparator shown in Fig. 3.6 are simulated, in simulation (a) only the input of the 1st gain stage is switched, whereas in simulation (b) the input of the 1st and 2nd stage is switched. The simulated output of version (a) is more or less undefined whereas the compensation technique works properly in case (b). The reason for this effect is the low gain of the single stages. In case (a) the gain of the 1st stage is not large enough to suppress the transient mismatch of the 2nd stage. Compared to auto zeroing the switched input technique works as background compensation and needs not to be applied before every decision. However, the static mismatch of

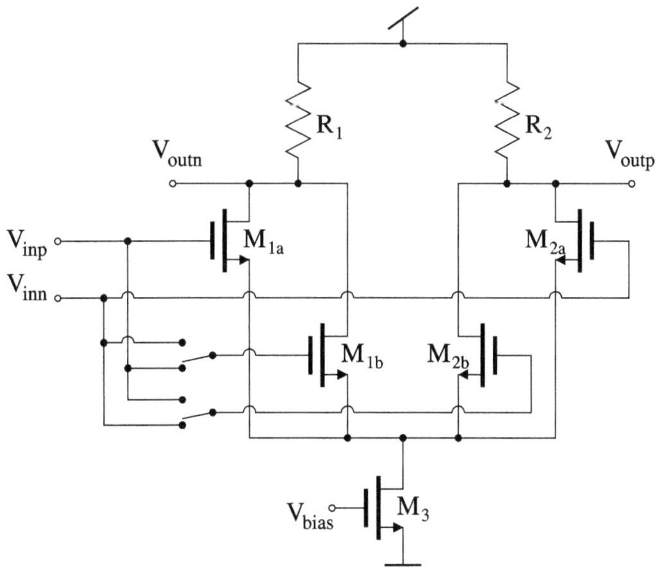

Fig. 3.12 Schematic of comparator with switched input devices

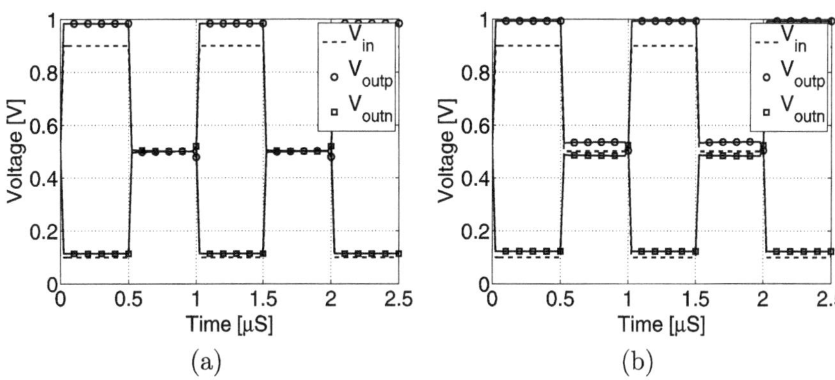

Fig. 3.13 Transient response of comparator with switched input in 1st stage (**a**) and with switched input in 1st and 2nd stage (**b**)

the comparator differs for both device combinations which has to be considered in some cases. Moreover glitches due to timing uncertainties and charge injection have to be taken into account.

Switched Bias Techniques Due to the identical physical background, flicker noise reduction techniques may also be applied to reduce the impact of charge trapping and hysteresis effects. The switched bias techniques shown above prevent the filling of traps as long as the clock period is shorter than the trapping time constants. As stated before the usage of switched bias techniques in linear analog applications imposes some drawbacks like switching noise. However, in switched capacitor circuits or comparators this concept may be implemented.

Summarizing the results so far it can be stated that switched capacitor amplifiers and comparators are the only building blocks noticeably affected by hysteresis effects. The consequences for analog-to-digital converters are discussed next.

3.2.3 Flash ADC

In n-bit flash A/D converters the analog input value is compared in parallel by 2^n comparators with 2^n equally spaced voltages, see Fig. 3.14(a). The reference voltage steps commonly are generated with a resistor ladder. The result of this parallel comparison is thermometer coded and has to be de-coded afterwards. Due to the exponentially growing number of comparators with increasing resolution, the corresponding power consumption and area practically limit the resolution to about 8 bit [91].

In case of transient mismatch the transition from zero to one in the comparator outputs is shifted. Incorrect comparator decisions can occur if the input value is close to a decision level. Therefore the resulting quantization error will be small, i.e. in worst case 0.5 LSB plus the maximum dynamic V_T shift. As long as the overall quantization error stays below 1 LSB the comparator performance is not affected significantly, i.e. the maximum V_T shift should not exceed 0.5 LSB as simple approximation. A quantitative example is shown in Fig. 3.14(b). The maximum

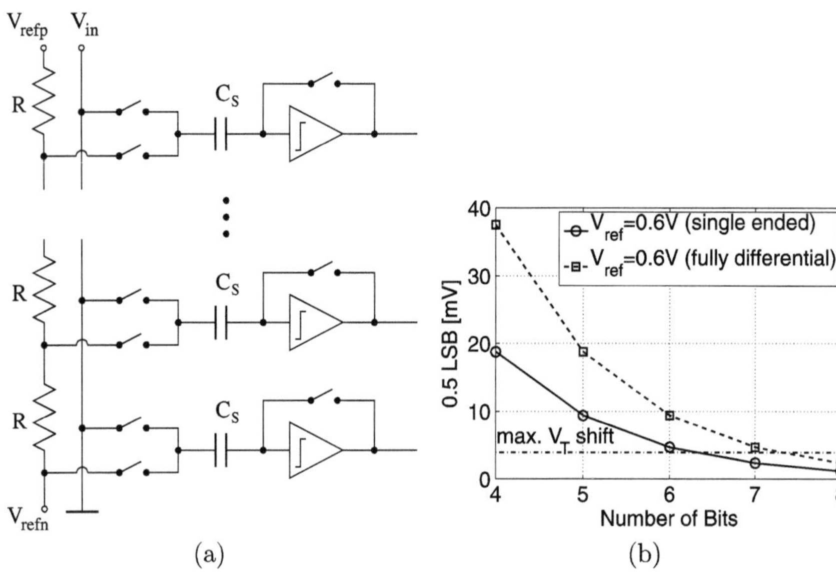

Fig. 3.14 Schematic of single ended flash ADC in switched capacitor technique (**a**) and 0.5 LSB value for varying number of bits (**b**)

measured V_T shift is compared to the 0.5 LSB value for varying number of bits, assuming a reference voltage $V_{ref} = V_{refp} - V_{refn}$ of 0.6 V. Fully differential and single ended ADC implementations are affected for resolutions higher than 7 and 6 bits, respectively.

In CMOS technologies flash converters can be easily implemented as switched capacitor circuit with implicit offset compensated sample and hold functionality [91], see Fig. 3.14(a). In this case the offset compensation drastically reduces the impact of transient mismatch, since the offset compensation is applied for every conversion. Moreover flash converters typically are used for very high conversion rates, i.e. at high speed compared to the (de)trapping time constants. Therefore the offset compensation works very accurate. Simulation results of a 6 bit flash ADC with $V_{ref} = 0.6$ V operated at 1 GS/s employing offset compensation shows no observable performance degradation for maximum dynamic V_T shifts below 10 mV.

3.2.4 Successive Approximation ADC

Successive approximation (SAR) converters are based on a binary search algorithm and work in a feedback loop including single bit A/D converter and n bit D/A converter. During the binary search the circuit halves the difference between the input signal and the DAC output in every step, i.e. the DAC output converges to the input value. Charge-redistribution DACs as shown in Fig. 3.15 are used commonly in CMOS technologies. For the sake of clarity the complementary capacitor network of the fully differential implementation is not explicitly shown in the schematic but indicated by a single capacitor at the positive comparator input. The conversion in the SAR-ADC is done in three steps. In the sample phase the input voltage V_{in} is stored at the bottom plates of the capacitor network while the top plates are grounded. Then the bottom plates are grounded (V_{refn} is assumed to be 0) and the top plates are left floating, changing the potential V_x at this node to $-V_{in}$. In the following conversion phase the bits are evaluated. First the bottom plate of the largest capacitor is

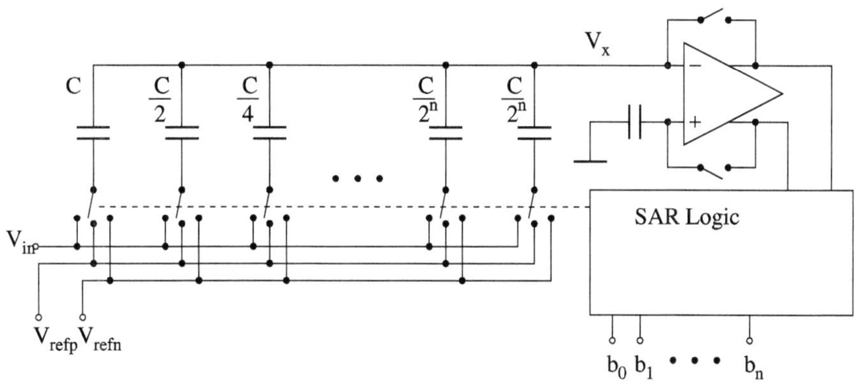

Fig. 3.15 Schematic of charge-redistribution based successive approximation ADC

switched to V_{refp}. Due to charge redistribution the potential at the comparator input will now be $-V_{in} + V_{refp}/2$ which is compared to V_{refn}. If the comparator output is positive the first bit is set to 0 and the voltage at the bottom plate of the capacitor is kept constant, otherwise the first bit is set to 1 and the voltage is switched back to V_{refn}. The following bits are evaluated in the same way.

3.2.4.1 Conversion Errors

Incorrect comparator decisions induced by transient mismatch result in a conversion error of at least one LSB [50]. Figure 3.16 shows a quantitative example. In this simulation a 12 bit SAR converter with a clock frequency of 50 MHz and a conversion rate of 3.6 MS/s is used. The comparator is offset-compensated and consists of three adjacent gain stages and differential latch as shown in Fig. 3.6. A single conversion cycle is shown in Fig. 3.16. The input value $-V_{in}$ is slightly higher than $7/8\,V_{refp}$. The comparator input voltage, i.e. the potential V_x at the top plates of the capacitor network, converges successively to zero. The first and second bit are set correctly. Switching the comparator input voltage from $-V_{in}$ to $-V_{in} + V_{refp}/2$ and further to $-V_{in} + 3/4V_{refp}$ is not critical since the residual input voltage is still about $-3/8V_{refp}$ and $-1/8V_{refp}$, respectively. In the third step the comparator input voltage is switched to a value close to zero. Transient V_T mismatch induces an incorrect comparator decision, the third bit is set faulty. The following bits are evaluated correctly, since no similar situation occurs.

Figure 3.17(a) shows the conversion error of the 12 bit SAR in dependence of the input value assuming a maximum V_T shift of 4 mV. For input voltages lower than $V_{ref}/2$ very few errors occur. For small input voltages the comparator input is never switched from sufficient large to very small values during the conversion, which would generate incorrect decisions. Input values around $V_{ref}/2$ and $3/4V_{ref}$ yield conversion errors around one LSB. The conversion error raises up to 3 LSB for further increasing input voltage. In this case several incorrect comparator decisions

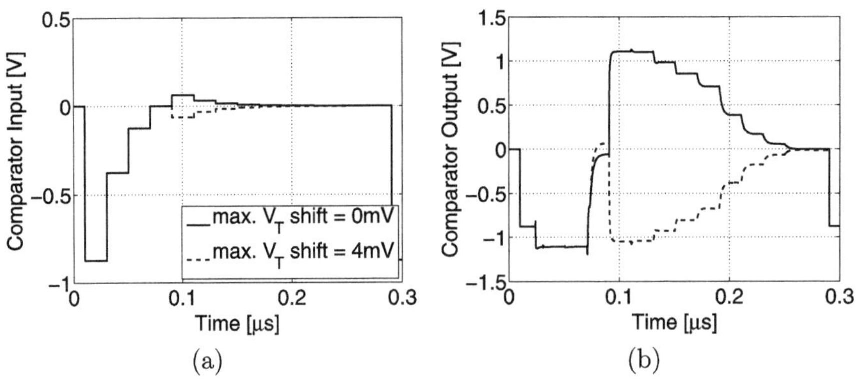

Fig. 3.16 Comparator input (**a**) and output (**b**) voltage during conversion with maximum V_T shift = 0 mV and maximum V_T shift = 4 mV

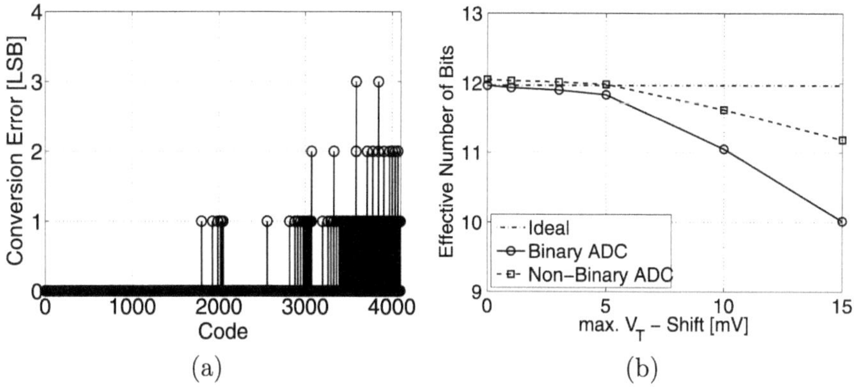

Fig. 3.17 Conversion error of 12 bit SAR ADC with maximum V_T shift = 4 mV (**a**) and effective resolution of binary and non-binary SAR-ADC for varying maximum V_T shift (**b**)

affect the result. Also the density of faulty codes increases with V_{in}. The correlation of the input signal to the conversion error leads to harmonic distortion in the output spectrum which again degrades the effective resolution, i.e. the effective number of bits (ENOBs) of the ADC drops. The reduction of ENOBs for the 12 bit SAR with rising maximum V_T shift is shown in Fig. 3.17(b).

3.2.4.2 Countermeasures

SAR converters are very sensitive to dynamic V_T variations. Thus countermeasures are required to enable high resolution SAR ADCs in case of transient mismatch.

Offset Compensation Charge redistribution based SAR converters are inherently offset compensated. The comparator offset is stored on the capacitor network during sampling and auto zeroed during conversion phase. However the offset compensation has to be applied before the conversion, otherwise the information stored on the capacitor network would be lost. Hence dynamic mismatch which is generated during the conversion is not compensated.

Comparators with switched inputs can be used, since the compensation also works during the conversion phase. The accuracy of this concept is limited by charge injection of the input switches into the capacitor network. A more efficient compensation is possible on architectural level.

Non-binary Search In [92] a SAR converter based on a non-binary search algorithm is proposed to relax the comparator and/or buffer settling and speed requirements. Non-binary means a basis smaller than 2 for the search algorithm, i.e. an input value can be represented by more than one digital code. This enables some kind of error correction since different ways through the search can lead to the same (decimal) value. Wrong comparator decisions can be corrected by this means. Figure 3.18 illustrates this concept. Using a basis smaller than 2 for the conversion

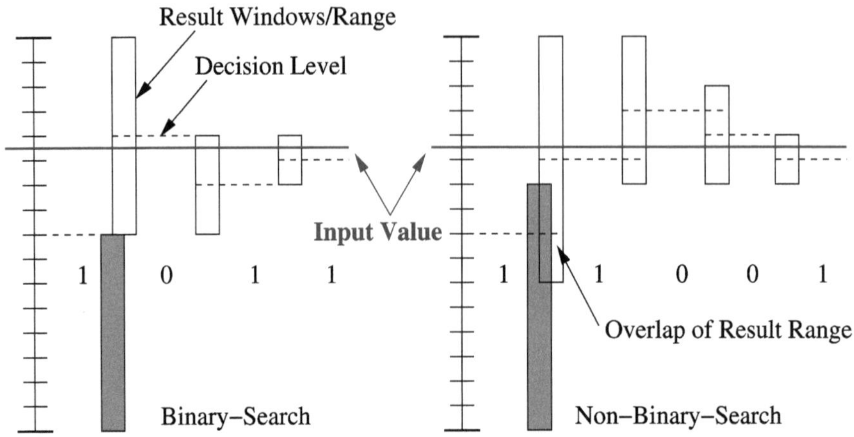

Fig. 3.18 Binary and non-binary successive approximation scheme

results in overlapping decision ranges, the algorithm contains redundancy. A one in the first comparator decision means that the input is larger than half of the reference plus some error tolerance. Obviously more clock cycles are required to achieve the same resolution. Since small DAC and comparator settling errors are now compensated the clock frequency can be increased and the overall conversion time stays constant or gets even smaller [92].

This error correction mechanism based on redundancy can also be used to correct errors caused by transient mismatch [66]. Figure 3.17(b) compares the simulated signal-to-noise ratio of the binary 12 bit SAR mentioned above with a non-binary implementation including 12% redundancy. For V_T shifts below 5 mV no serious degradation occurs, whereas the effective number of bits of the binary converter is reduced by about one for 10 mV and about two for 15 mV. The non-binary ADC is able to compensate all errors resulting from V_T shifts below 5 mV. In case of larger V_T shifts the resolution is much less degraded for the non-binary ADC. 12% redundancy implies two additional conversions, resulting in about 20% power overhead. The concept can be used to compensate for even larger V_T shifts by adjusting the amount of redundancy.

3.2.4.3 Implementation Aspects and Measurement Results

To prove the efficiency of the error correction mechanism with respect to transient mismatch a 12 bit SAR-ADC with non-binary search is implemented in an experimental 32 nm planar high-k metal-gate CMOS technology based on [67]. The block level schematic and the layout view of the implemented SAR-ADC test-vehicle is shown in Fig. 3.19. The ADC consists of input buffer, differential capacitor array, comparator and digital part. The input buffer is shared between the analog input and the reference signals. For the comparator 4 offset compensated gain stages are used. The differential capacitor array of the feedback DAC is driven by a thermometer

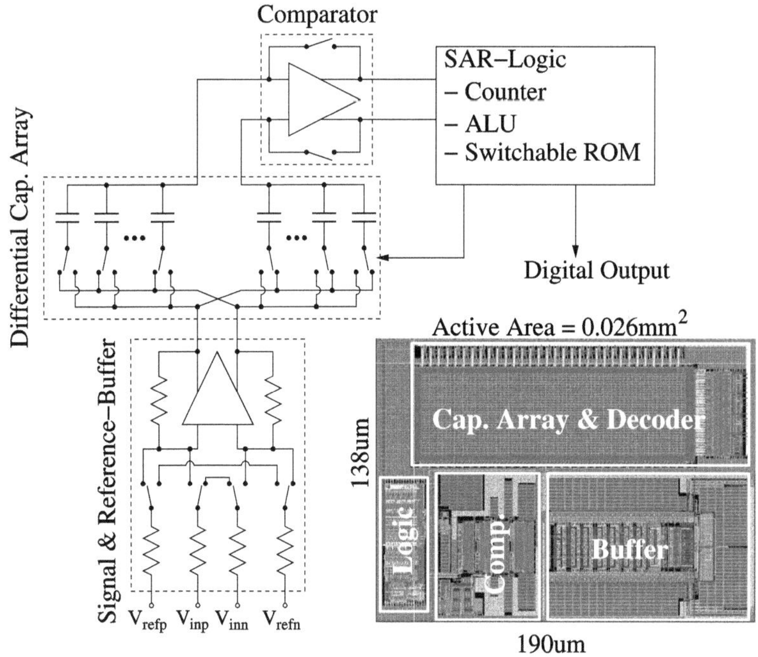

Fig. 3.19 Block level schematic and layout view of implemented 12 bit SAR-ADC

code for different reasons: a low differential non-linearity is inherently guaranteed, the power consumption of the reference buffer is reduced since no huge MSB capacitor needs to be driven and finally the problem of non-binary scaled capacitor values in the DAC can be avoided. The redundancy is not realized in the DAC but shifted to the digital part of the converter. The non-binary bit weights are stored in a ROM. During the conversion an arithmetic unit calculates the next DAC values based on the ROM content. The ROM content can be changed with an external control bit to change from non-binary to binary conversion. Adjusting the ROM content in principle also allows to match the amount of redundancy to the amount of V_T shift.

At nominal operating conditions (sampling frequency $f_{sample} = 5$ MS/s, analog input frequency $f_{ain} = 1.5$ MHz) the ADC achieves an effective resolution of more than 10 bits in non-binary mode. In this case the ADC draws about 10 mA from a 1 V supply. The overall current consumption is dominated by the input buffer which burns about 70% of the overall current. The effective resolution at nominal conditions is mainly determined by the capacitor mismatch and the thermal noise of the input buffer.

Figure 3.20 shows the measured effective resolution of the SAR-ADC for varying sampling frequencies in binary and non-binary mode. The ADC achieves more than 10 ENOBs over a wide frequency range (~1–10 MS/s) in non-binary mode. With further increasing sample rate the resolution is significantly reduced due to the speed limitations of input buffer and comparator that cause severe settling errors. However, compared to binary mode the drop in resolution occurs at much higher

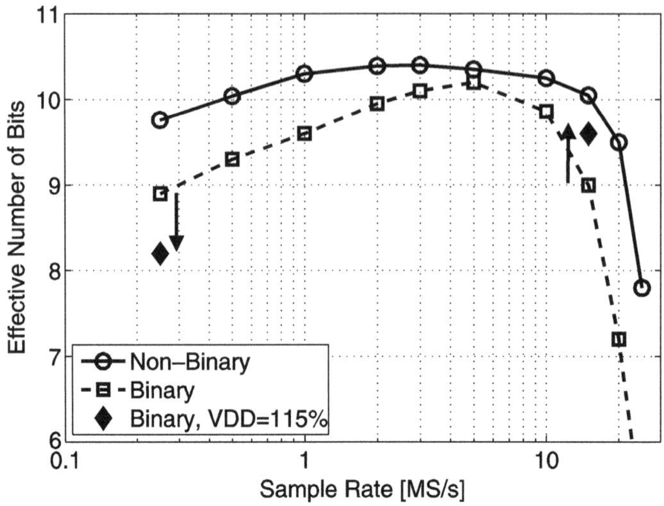

Fig. 3.20 Measured effective resolution of 32 nm 12 bit SAR-ADC for varying frequency in binary and non-binary mode

frequencies. This behavior shows that moderate settling errors can be corrected by the redundancy. The effective resolution in non-binary mode is also slightly degraded at very low sampling frequencies. This effect can be explained with different leakage current mechanisms that change the stored charge on the capacitor network, e.g. gate leakage at the comparator input devices or gate-induced drain leakage at the offset compensation switches. Obviously the amount of lost charge increases with decreasing sampling rate. It is noteworthy that conversion errors caused by leakage and lost charge can not be corrected by means of redundancy. In binary mode the degradation of effective resolution is considerably worse. In addition to leakage here also charge-trapping induced transient mismatch contributes to the degradation of ENOBs. Again the resolution decreases with decreasing sampling rate, since the amount of dynamic V_T-shift increases with stress time as long as there are still non occupied traps. This saturation is not observed in the measurements because the actual clock period (sampling period divided by number of conversion steps) is still too small.

This interpretation is supported by the behavior at higher supply voltages in binary mode. At high sampling rates the increased supply voltage reduces settling errors and improves resolution. However at low sampling rates an increased supply voltage results in higher leakage currents and hysteresis effects. The results confirm that error correction by means of non-binary search reduces the impact of transient mismatch significantly.

3.2.5 ΣΔ ADC

The impact of hysteresis effects on ΣΔ A/D converters is discussed next. To simplify matters the analysis is restricted to single bit converters. The basic concept of ΣΔ A/D converters is the use of oversampling and feedback to improve the resolution of a coarse quantizer [93]. Oversampling lowers the in-band quantization noise, since the total noise power is distributed over a wider band. The employment of feedback allows to shape the quantization noise out of the band of interest, whereas the signal is not affected. A digital decimation filter is required to remove the out of band noise.

The block level schematic of a first order single bit modulator is shown in Fig. 3.21. The signal is just delayed by the modulator, whereas the quantization noise is high-pass filtered. The corresponding implementation of the ΣΔ modulator as switched capacitor circuit is also shown in Fig. 3.21.

The blocks affected by transient mismatch are integrator and comparator. As shown in Sect. 3.2.2.1 the dynamic V_T shift causes a transient offset in the integrator OpAmp, which is comparable to $1/f$ noise in the output spectrum. Due to the low frequency components this additional noise is typically not shaped and therefore reduces the signal to noise ratio. Fully differential implementations do not suffer from this effect if mismatch of hysteresis parameters is neglected. However, the

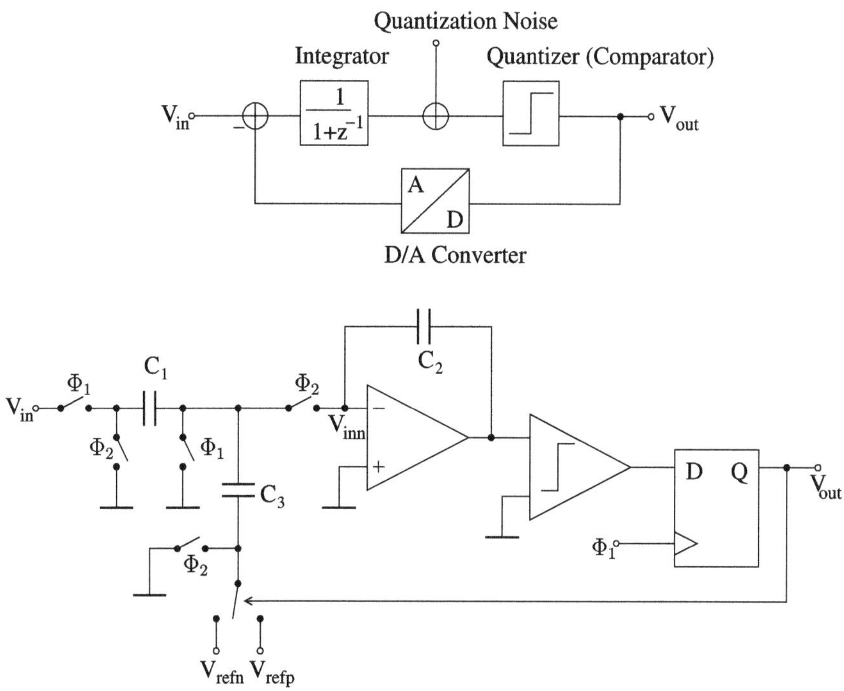

Fig. 3.21 Block level schematic of 1st order ΣΔ modulator and corresponding switched capacitor implementation

comparator is affected by transient mismatch in any case. Fortunately $\Sigma\Delta$ converters are quite robust against comparator errors: from system point of view incorrect comparator decisions can not be distinguished from quantization noise, they are generated in the same part of the loop. Consequently comparator errors are shaped in the same way as quantization noise, i.e. they are suppressed by the loop filter. Moreover incorrect comparator decisions only occur at small input values (compared to the common mode level). In this case the corresponding inaccuracy and quantization noise resulting from either decision is almost equal.

The degradation of $\Sigma\Delta$ converter performance due to dynamic V_T shifts is estimated with a simple approximation: the maximum dynamic V_T shift δV_{Tmax} is just added to the maximum quantization error $\Delta = 0.5$ LSB. The signal-to-noise ratio is then calculated using [94]

$$S = \frac{\Delta^2 2^2}{8}, \tag{3.3}$$

$$N_{\text{1st order}} = \frac{(\Delta + \delta V_{Tmax})^2 \pi^2}{36} \left(\frac{1}{\text{OSR}}\right)^3, \tag{3.4}$$

$$N_{\text{2nd order}} = \frac{(\Delta + \delta V_{Tmax})^2 \pi^4}{60} \left(\frac{1}{\text{OSR}}\right)^5, \tag{3.5}$$

with the oversampling ratio OSR. The estimated SNR degradation is shown in Fig. 3.22. As expected the converter performance is not very sensitive against dynamic V_T shifts. A maximum V_T shift of more than 10 mV reduces the SNR by less then 0.6 dB. Due to the simplification the estimated degradation of SNR is independent of oversampling ratio and modulator order. To verify this approach the

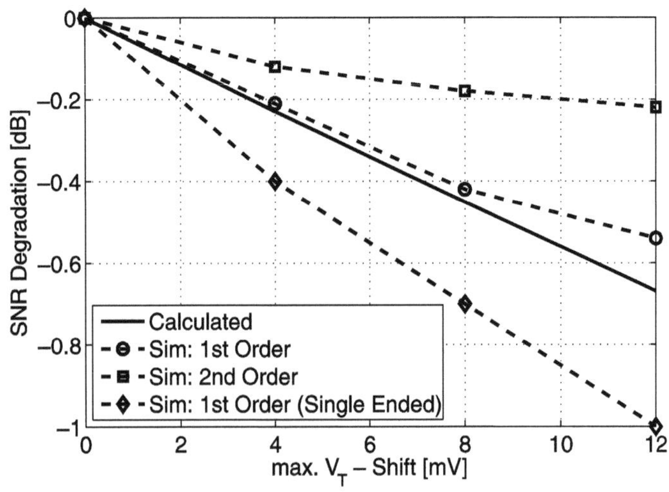

Fig. 3.22 Calculated and simulated SNR degradation of 1st order and 2nd order $\Sigma\Delta$ converter for varying maximum V_T shift

calculation is compared to simulated results, see Fig. 3.22. First and second order $\Sigma\Delta$ modulators are simulated using the scalable hysteresis model. The modulators are implemented as switched capacitor circuit, the oversampling ratio and bandwidth are 64 and 100 kHz, respectively. The simulated SNR degradation of the 1st order modulator is close to estimation. However, the SNR degradation of the 2nd order modulator is overestimated. The deviation is caused by the different probability of incorrect comparator decisions. Compared to the 2nd order modulator the amplitude steps at the comparator input of the 1st order modulator are larger. The comparator inputs signal crosses the decision level more often, the error probability is higher.

Taking also the transient offset of the single ended integrator into account, the SNR reduction is larger, up to 1 dB in this example. Nevertheless the impact of transient mismatch on $\Sigma\Delta$ converters is negligible in most cases, especially in fully differential, single bit, high order designs. The impact on multi bit $\Sigma\Delta$ converters should be evaluated in further investigations.

3.2.6 Conclusions on Transient V_T Shift

The results presented above show that charge trapping and hysteresis effects with maximum V_T shift in the range of $2 \ldots 3$ mV are no show stopper for analog and mixed-signal circuit design, even if high resolution is required. Nevertheless the hysteresis effect has to be considered in some cases: on building block level comparators are significantly affected. Experimental results prove the predicted degradation of bit error rate. Depending on the application auto zeroing and/or switched input comparators can be used to compensate for transient V_T mismatch. On system level hysteresis effects primarily can degrade the resolution of A/D converters following the principle of successive approximation. Introducing non-binary search algorithms is proposed as efficient countermeasure to compensate for incorrect comparator decisions. Further work should address statistical variations of hysteresis parameters and wear out effects. Wear out effects, i.e. degradation of hysteresis parameters and V_T stability during device lifetime is not considered here due to missing measurement data.

Chapter 4
Multi-Gate Related Design Aspects

In this chapter analog and mixed-signal circuit design aspects related to multi-gate specific device behavior are discussed. The main goal is to provide an early, technology oriented circuit assessment, focusing on different variation aspects. Based on a close link to device and technology the feasibility of analog and mixed-signal circuits in emerging FinFET technologies is proven. Performance differences and advantages compared to planar CMOS are quantified with examples. The assessment covers a broad range of analog and mixed-signal circuits, starting from basic biasing blocks, over OpAmps to D/A converter and PLL as mixed-signal examples. An outlook to RF design aspects is given by the discussion of 2 GHz VCOs and LNAs. The final sections cover further design aspects related to the SOI FinFET device structure.

For all circuit simulations shown here a dedicated FinFET compact model is used [95]. The measurement results are obtained from FinFET devices as presented in Chap. 2 which are also used for model parameter extraction.

4.1 Biasing Circuits

Biasing circuits are key analog building blocks intended to generate and distribute stable reference currents and voltages, robust against variations of supply voltage and device parameters. Although also bandgap voltage reference circuits fit into this definition, they are covered in a separate section, since additional aspects such as p-n junctions and OpAmp design have to be considered.

4.1.1 Matching Optimized Current Mirrors

The current mirror is one of the most frequently used building blocks in analog circuits. Roughly speaking current mirrors are used to generate replica currents representing exact multiples of the input current, see Fig. 4.1(a). Deviations from ideal

M. Fulde, *Variation Aware Analog and Mixed-Signal Circuit Design in Emerging Multi-Gate CMOS Technologies,* Springer Series in Advanced Microelectronics 28,
DOI 10.1007/978-90-481-3280-5_4, © Springer Science+Business Media B.V. 2010

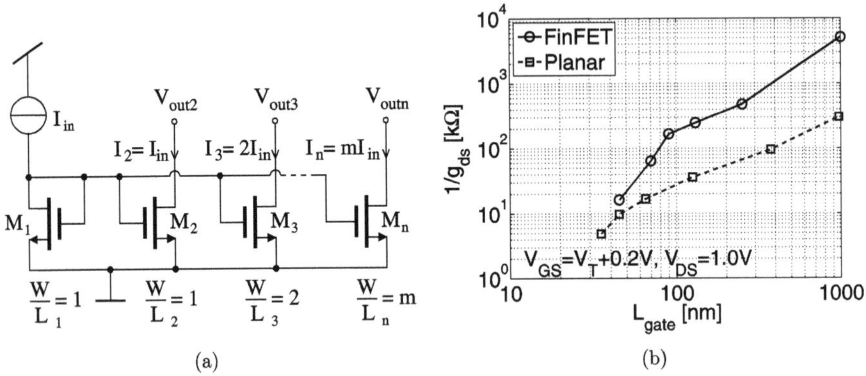

Fig. 4.1 Schematic of exemplary nFET current mirror with multiple outputs (**a**) and comparison of FinFET and planar output impedance for varying gate length (**b**)

current mirror behavior are caused by finite output impedance and device parameter mismatch in DC case. From AC perspective the constant bias voltage at the common gate is affected by output voltage variations which are AC coupled via the gate-drain capacitance. Thus, the devices should feature high output impedance and good matching at low area, i.e. low parasitic capacitances.

Obviously the low g_{ds} of FinFETs directly corresponds to an improved output impedance resulting in lower sensitivity against output voltage variations, see Fig. 4.1(b). The advantageous matching behavior can be used to decrease area and improve AC performance compared to planar implementations.

As mentioned in Chap. 2 the matching of FinFETs is affected by several layout dependent effects. A current mirror test-structure with multiple outputs as shown in Fig. 4.1(a) is used to compare different FinFET specific layout styles in order to derive matching optimized layout guidelines [66].

The "golden rules" for good matching in analog layout can be summarized in few catchwords: symmetry, unit devices & cells and regular environment. Following these guidelines several different basic layout styles are possible, as illustrated in Fig. 4.2:

- *Shared source/drain areas* (A): Each current mirror device is splitted into several unit devices with equal fin count and merged source/drain landing pads. Dummy gates are placed to ensure regularity. The current flows in anti-parallel directions.
- *Shared source/drain areas with dummy fins* (B): Same layout as (A) with additional dummy fins to improve regularity and to avoid contribution of "outer" fins that do not match "inner" fins.
- *Single source area* (C): The current mirror devices are splitted into several unit devices. All devices share a common source landing pad, i.e. gate misalignment affects all devices in the same way. The current flows in only one direction. Dummy devices improve regularity.

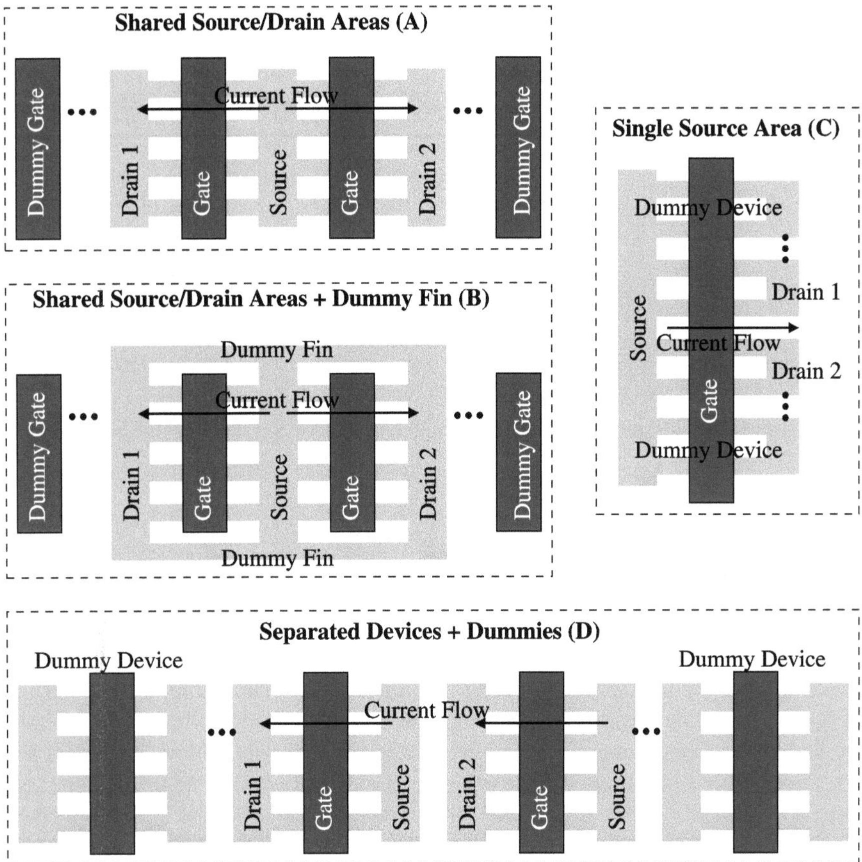

Fig. 4.2 Different layout styles for optimized matching

- *Separated unit devices* (D): Again the current mirror devices are splitted into several unit devices. The unit devices are separated, source/drain landing pads are not shared. Regularity is ensured by dummy devices.

In addition to the layout style itself also the amount of fins per unit device at constant overall device width impacts matching, because the relation of outer to inner fins changes. Several n- and p-type FinFET current mirrors with different layout style and varying amount of fins per unit device (finger) are compared. To enable a fair comparison the mismatch induced by random dopant fluctuations needs to be comparable. Thus, the active device area, i.e. the gate length and overall width is kept constant. The current mirrors are characterized at different input current levels representing low and high gate overdrive voltage ($V_{GS} - V_T = V_{ov} = 0.2$ V and 0.4 V). Matching performance is benchmarked by means of the relative current mismatch from input to output $\Delta I/I$ which is measured over a whole 200 mm wafer

Table 4.1 Measured standard deviation of current mismatch for n- and p-type FinFET current mirrors at low and high gate overdrive

Style	Fins/Finger	Area	nFinFET		pFinFET	
			$V_{ov} = 0.2$ V	$V_{ov} = 0.4$ V	$V_{ov} = 0.2$ V	$V_{ov} = 0.4$ V
A	8	1	8.43%	3.43%	9.28%	5.03%
A	4	1.07	9.51%	5.54%	11.12%	6.82%
A	2	1.14	13.05%	9.08%	14.92%	9.47%
B	8 + dummy	1.12	6.57%	3.09%	7.35%	4.61%
C	8	0.98	5.72%	3.20%	7.02%	4.13%
D	8	1.3	4.96%	2.13%	4.53%	1.83%

(>50 samples). The relative current mismatch $\Delta I/I$ of the n-th output is defined as

$$\frac{\Delta I}{I} = \frac{I_{in} - I_{out_n}/m_n}{I_{in}} \tag{4.1}$$

with the current mirror factor m_n. The mismatch of 4 outputs with m factors from 1 to 8 is averaged to increase the overall number of samples contributing to the statistics. To eliminate systematic errors not the absolute value but the standard deviation of current mismatch serves as metric. Table 4.1 summarizes the measurement results and compares the area overhead referred to the shared source/drain layout (A). As expected, matching improves with increasing overdrive voltage for all current mirror variations. Comparing the different realizations of layout option (A) with 2 fins per finger (only outer fins), 4 fins per finger (inner fins = outer fins) and 8 fins per finger (inner fins ≫ outer fins) shows, that a high amount of fins per finger is beneficial for matching. The main reason is that the process control over the fin shape is worse for the outer fins. In addition the area efficiency decreases with decreasing fin, i.e. increasing finger count due to the overhead resulting from gate connection and gate overlap beyond the fins. The fraction of gate connection and gate overlap compared to the overall height increases with decreasing fin count, see Fig. 4.2. Adding dummy fins to enhance regularity as done in layout option (B) improves matching but requires about 12% more area. Even better matching is achieved with layout option (C). In addition to the high regularity in this case also the direction of current flow is equal for all devices, minimizing gate misalignment induced variations. Also from area perspective this option is very attractive since the overlap resulting from gate connection and gate overlap is minimized in this case. The best matching performance is achieved with option (D) achieving the highest regularity. However, the high linearity is paid with a large area overhead of about 30% in this example.

4.1.2 Current Reference Circuits

Analog and mixed-signal circuit functionality relies on precise and robust reference and bias currents, where the required level of accuracy and robustness depends on

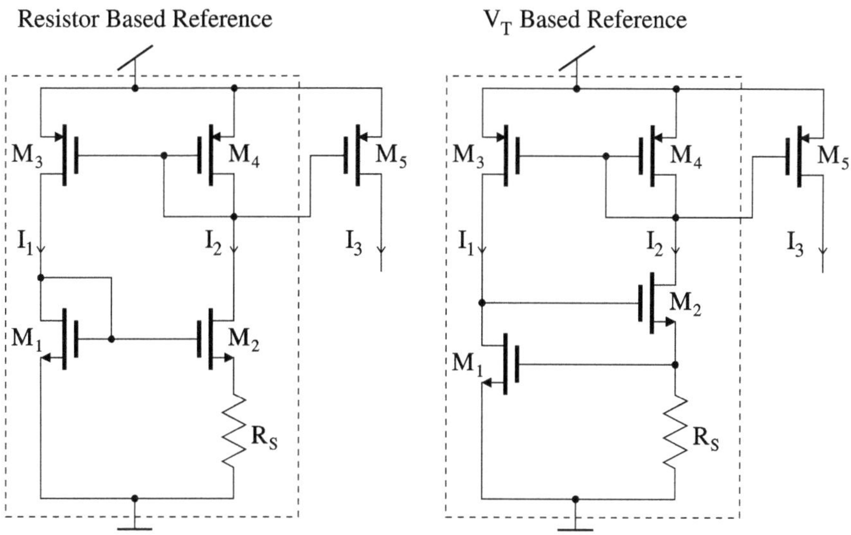

Fig. 4.3 Resistor (*left*) and threshold voltage (*right*) based current reference circuit, the *dashed line* indicates the core of the circuit. The W/L ratios of M_3 and M_4 are equal

the actual application. Temperature independent and accurate (in terms of absolute value) reference currents typically have to be derived from a high precision bandgap voltage reference circuit [96]. The design of bandgap references in FinFET technology is discussed later. However, for some applications the temperature dependence of the reference current is of minor importance, e.g. in local biasing networks of OpAmps or comparators. In this case simple reference circuits are used to generate bias currents which are insensitive against supply and output voltage variations. Consequently the figures-of-merit to assess and compare the circuit performance are the sensitivity against supply and output voltage:

$$R_{DD} \equiv \frac{\Delta V_{DD}}{\Delta I_{out}}; \qquad R_{out} \equiv \frac{\Delta V_{out}}{\Delta I_{out}}. \tag{4.2}$$

Figure 4.3 shows two commonly used current references, based on the concept of "self-biasing" or bootstrapping. The main idea to eliminate the supply voltage dependence is to derive the reference current ($= I_1$) from the output current ($= I_2$) and vice versa. With ideal current sources, i.e. zero output conductance in this way the reference current is independent of the supply voltage. The output current is scaled and distributed with a further current mirror (M_4:M_5).

4.1.2.1 Resistor Based Reference

Assuming that all devices are operated in saturation, $R_S = 0$ and ignoring channel length modulation ($g_{ds} = 0$) the resistor based reference is only defined by the two

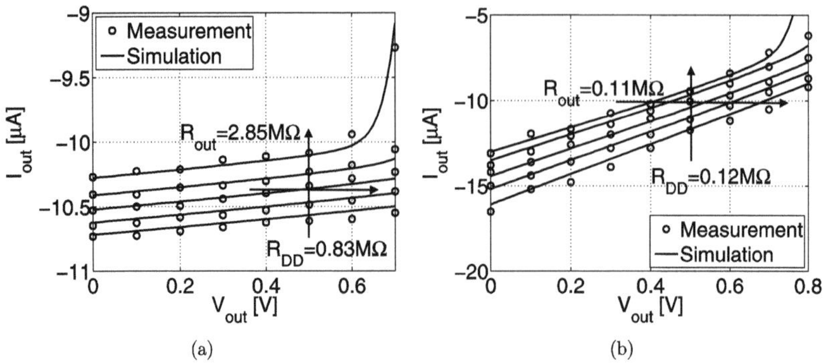

Fig. 4.4 Measured and simulated I_{out} versus V_{out} for V_{DD} from 0.8 V to 1.2 V for FinFET (**a**) and planar bulk resistor based current reference (**b**)

current mirrors and one single equation: $I_1 = I_2$. Thus I_1 and I_2 are independent of V_{DD}. As this equation is fulfilled for any current level, the resistor R_S is added and $(W/L)_2$ is chosen k times $(W/L)_1$ to define an unique current. The reference current is then given as

$$I_2 = \frac{1}{\beta_1} \frac{1}{R_S^2} \left(1 - \frac{1}{\sqrt{k}}\right)^2. \tag{4.3}$$

Since the reference current is distributed by the current mirror M_4:M_5 the output resistance R_{out} is just given by the g_{ds} of M_5. As shown in Fig. 4.1(b) the low FinFET output conductance directly corresponds to higher robustness against variations of V_{out}. The dependence of the reference current on the supply voltage can be calculated using a small signal equivalent circuit [45] and is approximately given by

$$\frac{V_{DD}}{I_{out}} \approx \frac{1}{g_{ds3}} (g_{m1} R_S + R_S g_{ds3} - 1). \tag{4.4}$$

To achieve a high R_{DD} the output conductance of M_3 should be as low as possible. Again, the improved FinFET g_{ds} yields better circuit performance.

To verify the benefits on silicon the resistor based current reference is implemented and characterized in FinFET and planar technology. Both versions feature n- and pFET gate lengths around 250 nm and approximately consume the same area. The layout is based on shared source/drain areas. Measured and simulated results are shown in Fig. 4.4. The output current is depicted versus the output voltage for supply voltages from 0.8 V to 1.2 V. R_{DD} and R_{out} are measured at $V_{DD} = 1.0$ V and $V_{out} = 0.5$ V respectively. As expected the FinFET reference is much less sensitive against variations of V_{out} and V_{DD}.

4.1.2.2 Threshold Voltage Based Reference

The self-biasing in the V_T based reference works slightly different. The current mirror M_3:M_4 ensures $I_1 = I_2$. I_1 determines the gate source voltage of M_1. This voltage drop across R_S again defines I_2, establishing an equilibrium point. The reference current can be written as [62]

$$I_1 = \frac{V_{T1}}{R_S} + \frac{1}{\beta_1 R_S^2} + \frac{1}{R_S}\sqrt{\frac{2V_{T1}}{\beta_1 R_S} + \frac{1}{\beta_1^2 R_S^2}}. \tag{4.5}$$

The distribution of the output current is carried out again with the current mirror M_4:M_5. Consequently the dependence of output current on output voltage is comparable to the resistor based circuit. The dependence on supply voltage can be approximated with

$$\frac{V_{DD}}{I_{out}} \approx \frac{1}{g_{ds1}} + \frac{1}{g_{ds3}} - \frac{g_{m1}}{g_{ds1}} R_S. \tag{4.6}$$

Again, high R_{DD} is achieved with low output conductance of M_3. Similar to the resistor based reference the circuit is implemented in FinFET and planar technology using n- and pFET gate lengths around 250 nm and shared source/drain layout style. Figure 4.5 compares measured and simulated dependence of the reference current on the supply voltage for FinFET and planar implementation. Due to the low FinFET g_{ds} the sensitivity against supply voltage variations is improved by more than a factor of 7.

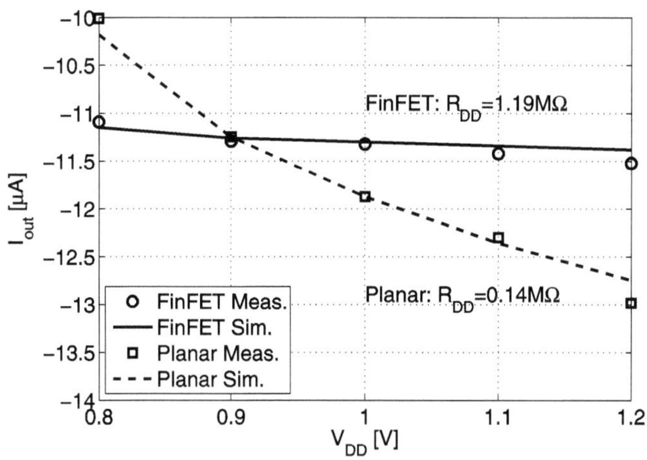

Fig. 4.5 Measured (*symbols*) and simulated (*lines*) dependence of V_T based reference current on supply voltage for planar and FinFET implementation

4.2 Operational Amplifiers

The design of operational amplifiers (OpAmps) is discussed in this section. The OpAmp is one of the most versatile and important building blocks in analog and mixed-signal circuits. An OpAmp is defined as high gain amplifier or controlled voltage source and is commonly used in negative feedback configurations. Assuming high OpAmp gain the closed loop transfer function is independent of the absolute value of OpAmp gain, enabling high accuracy buffer, amplifier or integrator functionality, defined by the feedback circuitry [62]. To simplify matters buffered operational amplifiers and un-buffered operational transconductance amplifiers are not separated since all considerations are focused on the gain stages featuring high output resistance. For the same reason only single ended OpAmps are considered. Adding a low impedance output buffer or common mode feedback does not change the conclusions given here.

Obviously sufficiently large open loop gain is a primary requirement for OpAmps. In this context the low intrinsic gain of scaled planar CMOS is a serious concern. Besides open loop gain the main design parameters are

- Gain-Bandwidth-Product (GBW), also called unity gain frequency,
- Phase margin,
- Slew rate and settling behavior,
- Common mode and power supply rejection ratio (CMRR, PSRR),
- Noise,
- Power consumption,
- Area.

The widely used two stage, Miller compensated OpAmp serves as design example in this section. As illustrated in Fig. 4.6 it basically consists of a differential input

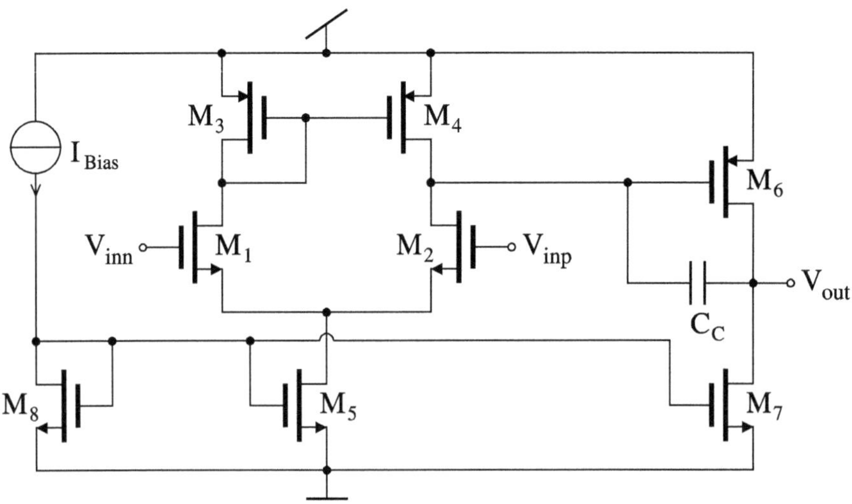

Fig. 4.6 Schematic of two stage, Miller compensated operational transconductance amplifier

stage with current mirror load followed by a second common source stage. Both stages contribute to open loop gain with their high impedance output. Therefore a compensation capacitance C_C is needed to separate the poles of both stages and guarantee stability. A comprehensive analytical description of the two stage Miller OpAmp regarding pole splitting, pole frequencies, bandwidth and noise is given e.g. in [61]. The important design equations for open loop voltage gain A_0, Slew-Rate (SR), Gain-Bandwidth-Product (GBW) and non-dominant pole frequency f_{nd} are

$$A_0 = \frac{g_{m1/2}}{g_{ds2} + g_{ds4}} \cdot \frac{g_{m6}}{g_{ds6} + g_{ds7}}; \qquad \mathrm{SR} = \frac{I_5}{C_C}, \tag{4.7}$$

$$\mathrm{GBW} = \frac{g_{m1/2}}{2\pi C_C}; \qquad f_{nd} = \frac{g_{m6}}{2\pi C_L} \cdot \frac{1}{1 + \frac{C_{n1}}{C_C}} \tag{4.8}$$

where C_L is the load capacitance and C_{n1} the parasitic capacitance at the output of the first stage. Besides (4.7) and (4.8) two additional constraints have to be considered: to achieve reasonable phase margin, i.e. stability in feedback configuration, the non-dominant pole frequency has to be $2\dots3$ times larger than GBW. The equations for pole frequencies are only valid if $C_{n1} < C_C < C_L$. Typically a factor $2\dots3$ is chosen, i.e. $C_L = 2\dots3C_C$; $C_C = 2\dots3C_{n1}$. This results in $g_{m6} \approx 5\dots10g_{m1/2}$, making the current in the second stage much larger than in the input stage.

4.2.1 Gain-Bandwidth-Power Trade-off

As shown in Chap. 2 the small signal parameters of planar and FinFET devices significantly differ. The consequences of these differences on the gain-bandwidth-power trade-off in OpAmp design is discussed next. Obviously the high intrinsic gain g_m/g_{ds} of FinFETs results in improved open loop gain comparing planar and FinFET OpAmp implementations with equal specifications and transistor dimensions. Assuming that the second stage consumes m times the current of the first stage the open loop gain of a two stage amplifier can be approximated as

$$A_0 \approx \frac{1}{\sqrt{m}} \left(\frac{g_m}{2g_{ds}} \right)^2. \tag{4.9}$$

Figure 4.7 shows the achievable open loop gain for a planar and FinFET two stage OpAmp with an m factor of 5. The power consumption of FinFET and planar OpAmp is almost equal: the slightly lowered FinFET g_m per unit width is compensated by the increased efficiency g_m/I_D, i.e. less drain current is needed to obtain the same g_m.

The consequences of the higher open loop gain of the FinFET implementation depend on the actual application and perspective. As mentioned above the OpAmp open loop gain determines the accuracy achievable in closed loop configuration [62]. Hence, one straightforward option is to benefit from improved accuracy on "system"

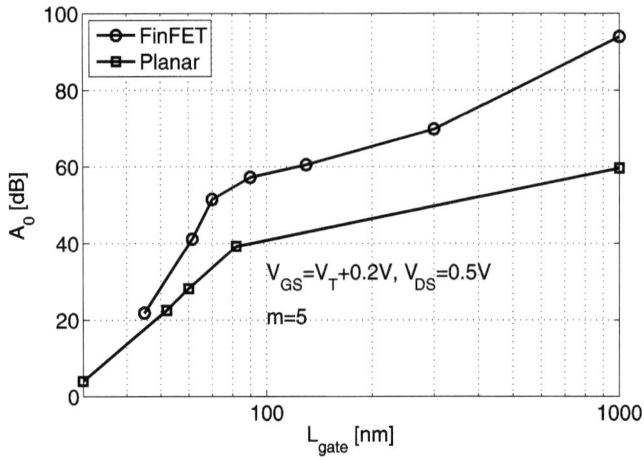

Fig. 4.7 Approximated DC voltage gain of planar and FinFET two stage Miller OpAmp

level. However, if the FinFET implementation just meets the open loop gain specification the planar design has to be tuned. Principally there are two ways to boost open loop gain: increasing the channel length, i.e. increasing g_m/g_{ds} or changing the topology. Improving the open loop gain by increasing the channel lengths is limited twofold. The intrinsic gain saturates with increasing gate lengths due to the HALO induced residual DIBL as explained in Chap. 2, limiting the maximum gain to about 60 dB for the planar reference technology. In addition the gate width has to be increased by the same amount to keep g_m constant. Increasing L and W results in quadratically raising device area and capacitance. If feedback is applied, the additional pole at the input is shifted to lower frequencies which may degrade phase margin and cause stability issues. If the required gain is not achievable with larger gate length the only solution is to change the topology, e.g. by adding a cascode or another gain stage [96]. Changing the topology for higher gain however implies a more or less serious power penalty. The only reasonable option for cascodes in low supply voltage scenarios is a folded cascode input stage. Besides the overhead induced by the generation of the required additional bias voltages the bias current of the first stage has to be increased by about 50% [45] yielding an overall power increase of about $(1.5+m)/(1+m)$. Adding a third gain stage introduces an additional pole which causes even more power overhead: To ensure stability the frequency of the third pole has to be about 5 times larger than the GBW, the overall current has to be almost doubled [96].

The FinFET perspective is different: if the high gain of the FinFET implementation is not required by the application, it can be traded in some cases against higher bandwidth or lower power consumption by optimizing bias point and dimensions of M_6. The optimization is based on the fact that M_6 can be operated with higher efficiency g_m/I_D if less gain contribution is required, i.e. if the channel length can be reduced. The required g_m of transistor M_6 is determined by the non-dominant pole frequency which has to be about $2\ldots3$ times higher than GBW for reasonable phase margin. The minimum gate length is given by the gain requirements. Hence there are two (dependent) design parameters left: the gate width and the overdrive voltage $V_{GS} - V_T$. For a power efficient design the overdrive voltage is chosen as small as possible, since the efficiency g_m/I_D degrades with increasing overdrive, see Chap. 2. This is limited by stability concerns: lowering $V_{GS} - V_T$ implicates higher channel width to keep g_m constant and consequently results in higher capacitance C_{n1} which is dominated by the gate source capacitance of M_6. Since the relation of C_{n1}/C_C affects the phase margin, there is an upper limit for the width of M_6 and a lower limit of $V_{GS} - V_T$ which is given by $C_{n1} \approx 2/3W_6L_6C_{ox}$. Compared to planar, the minimum FinFET gate length to reach the gain specification is smaller. Consequently, FinFETs can be biased with smaller overdrive and higher efficiency but same g_m, device area and phase margin. The better efficiency yields lower power consumption since the overall current is typically determined by the second stage. In a similar way also the gain-bandwidth of FinFET OpAmps can be improved for equal power consumption with respect to planar: the power saved by the increased efficiency is spend partly on the input and the output stage: a small fraction is used to enlarge g_{m1}, the remaining part is used to increase g_{m6} to retain stability. It is noteworthy that this optimization procedure is not applicable to the input stage, since decreasing $V_{GS} - V_T$ results in worse matching and noise performance [61].

4.2.2 Design Example

To quantify the statements given above, different versions of planar and FinFET Miller OpAmps have been designed according to the following specifications:

- $V_{DD} = 1.0$ V,
- $A_0 > 50$ dB,
- GBW > 10 MHz,
- $C_L = 5$ pF.

For a fair comparison of the different implementations the ratio of GBW per power consumption serves as figure-of-merit. The results are summarized in Table 4.2.

Table 4.2 Comparison of planar and FinFET Miller OpAmp performance featuring different channel lengths

	Planar	FinFET	FinFET	FinFET	Planar folded cascode
	$3L_{min}$	$3L_{min}$	$1.5L_{min}$	$1.5L_{min}$	$3L_{min}$
GBW [MHz]	10.8	10.6	**14.7**	10.4	10.6
P [μW]	65.2	66.4	68.8	**50.9**	75.9
A_0 [dB]	49.1	**66.3**	48.3	48.6	64.2
FOM [MHz/μW]	0.166	0.160	0.213	0.204	0.140
FOM improvement	±0	−3%	**+29%**	**+23%**	−18%

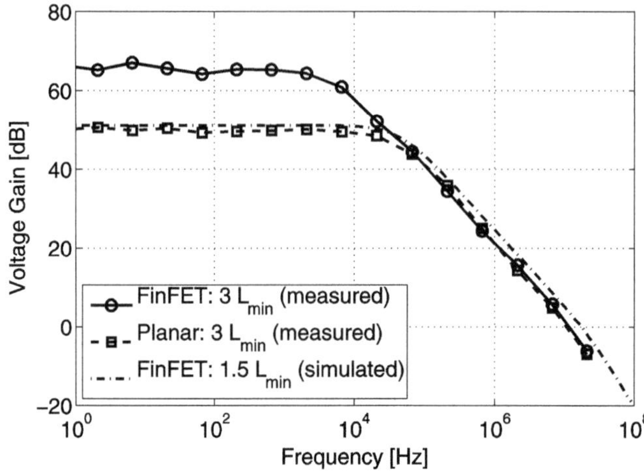

Fig. 4.8 Measured and simulated bode plots of planar and FinFET Miller OpAmp

Starting point is a planar reference design which almost meets the gain and bandwidth specifications. The input devices of both gain stages within this OpAmp have a gate length of about $3L_{min}$. Using the same gate lengths in the FinFET version yields an amplifier with a significantly higher open loop gain of about 66 dB. The open loop gain can be traded against higher GBW or lower power consumption by using shorter channel lengths in the second stage. To achieve the 50 dB gain a channel length of only $1.5L_{min}$ is required. As explained above the devices can be biased at lower overdrive. Optimization for bandwidth yields about 40% improvement at constant power consumption. Otherwise keeping the GBW constant enables a reduction of power consumption by about 25%.

In case of higher open loop gain specifications (e.g. 60 dB), the planar version has to be modified. Simulations with the planar reference technology show that achieving considerably more than 50 dB gain with reasonable power consumption is very demanding. For the specifications used here a Miller compensated OpAmp with folded cascode input stage and additional common source stage turned out to be the most power efficient alternative. The power overhead is about 15% in this case. These results confirm that the low intrinsic gain of scaled planar CMOS is one of the major challenges in analog design.

To verify the models and the simulation results, the $3L_{min}$ OpAmp versions are realized on planar and FinFET test chips. Figure 4.8 proves that the experimental results are close to the expectations in terms of gain and bandwidth. Also the power consumption is close to simulation.

4.2.3 Common Mode and Power Supply Rejection Ratio

The beneficial FinFET g_{ds} also reduces the sensitivity of OpAmps against common mode and power supply variations. A major source of finite common mode rejection ratio CMRR is the output impedance of the tail current source M_5. Especially in low-voltage applications low g_{ds} is crucial since commonly no voltage headroom is left for the implementation of cascoded current sources. According to [61] the CMRR of the two stage Miller OpAmp at low frequencies is given by

$$\mathrm{CMRR} \approx \frac{2 g_{m1} g_{m3}}{g_{ds5} g_{ds1}}. \tag{4.10}$$

Consequently reduced g_{ds} quadratically improves CMRR. In a similar way the power supply rejection ratio PSRR is improved. Quantitatively the advantage of employing FinFET devices can be approximated at low frequencies [62] with

$$\mathrm{PSRR_{VDD}} \approx \frac{g_{m1} g_{m6}}{g_{ds6}(g_{ds2} + g_{ds4})}; \qquad \mathrm{PSRR_{VSS}} \approx \frac{g_{m1} g_{m6}}{g_{m7}(g_{ds2} + g_{ds4})}. \tag{4.11}$$

4.3 Bandgap Reference Circuits

Bandgap reference circuits (BGRs) are key analog building blocks for system-on-chip applications, e.g. as part of power management units or A/D converters. The feasibility of bandgap references with sufficient performance is a strong requirement for the successful introduction of any multi-gate CMOS technology. The main objective of a voltage reference circuit is to generate a stable bias voltage, robust against variations of supply voltage, temperature and process/device parameters. Adding two well defined voltages featuring opposite temperature coefficients with proper weighting is a common approach to generate a temperature independent reference voltage [97]. In the so called bandgap reference a forward p-n diode voltage V_D provides the negative temperature coefficient (NTC).[1] The temperature dependence is defined by the diode saturation current and is hardly affected by process variations [98]. The positive temperature coefficient (PTC) is obtained from the difference of two forward diode voltages, where the current through both diodes is equal but the area differs by a factor m. Due to the exponential behavior of the diode current $I_D = I_S \exp[V_D/(n V_{the})]$ the absolute value of the saturation current cancels out and the voltage difference is proportional to the thermal voltage V_{the}.[2] The implementation of bandgap references in multi-gate CMOS technologies is challenging for mainly two reasons: The low supply voltage of about 1 V strongly limits the voltage headroom of stacked devices. In addition, the basic elements of bandgap references, p-n junctions are not available in SOI technologies. This section presents design aspects and measurement results to prove the feasibility of low voltage bandgap reference circuits in multi-gate technologies. The main figures-of-merit to assess the bandgap performance are:

---

[1] NTC of $V_D \approx -1.8$ mV/°C at room temperature.

[2] PTC of $\Delta V_D \approx V_{the} \ln(m) = (k_B T/q) \ln(m)$.

- accuracy of the absolute value of the reference voltage V_{ref},
- power supply rejection ratio (PSRR) defined as $\partial V_{ref}/\partial V_{DD}$ and
- the temperature coefficient (TC) defined as $[V_{max}(T) - V_{min}(T)]/\Delta T$.

4.3.1 Gated p-i-n Diodes

In SOI technologies "intrinsic" p-n junctions are not available without additional process steps, therefore gated p-i-n diodes are used here instead. Gated p-i-n diodes consist of standard FinFETs where the doping of one of the source/drain regions is inverted, see Fig. 4.9. The gate is required for two reasons: in todays CMOS technologies typically silicide layers (e.g. NiSi) are used to reduce the contact resistance between silicon source/drain region and metal (e.g. tungsten) contact plug as shown in Fig. 4.9. The gate serves as silicide blocking layer, preventing a short of the low ohmic silicide on top of the n^+ and p^+ regions. Additionally the gate ensures a certain distance between the n^+ and p^+ regions which avoids band-to-band tunneling currents between anode and cathode that would distort the drift-diffusion dominated I–V characteristics of the diode. The undoped fins form the intrinsic region in between. The gate of the device is connected to the n^+ cathode.

To derive a model for circuit simulation several gated p-i-n diodes have been characterized stand alone. Measured I–V characteristics of a p-i-n diode with a gate length of 300 nm and 32 fins in parallel at different temperatures are shown in Fig. 4.10. In general, the electrical characteristics is similar to standard p-n junctions. However, the transition from the generation/recombination (G/R) region at

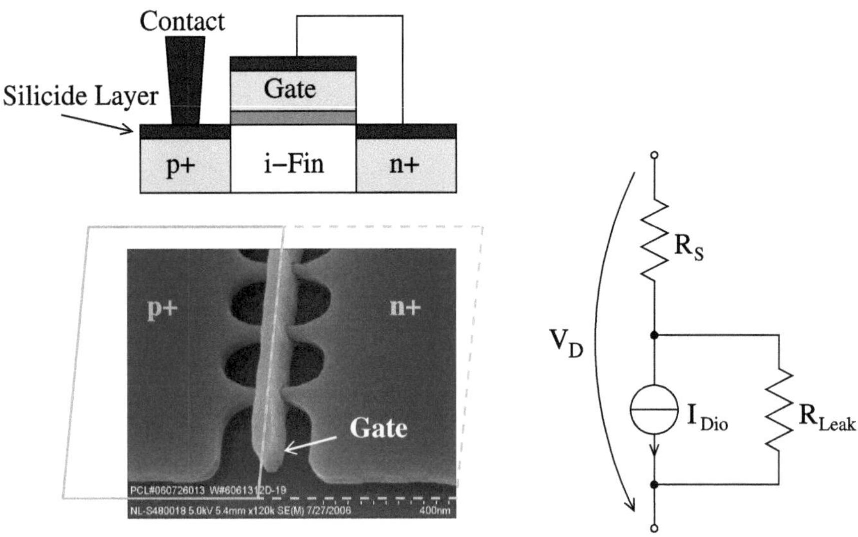

Fig. 4.9 Schematic cross-section, TEM picture and model of gated p-i-n diode

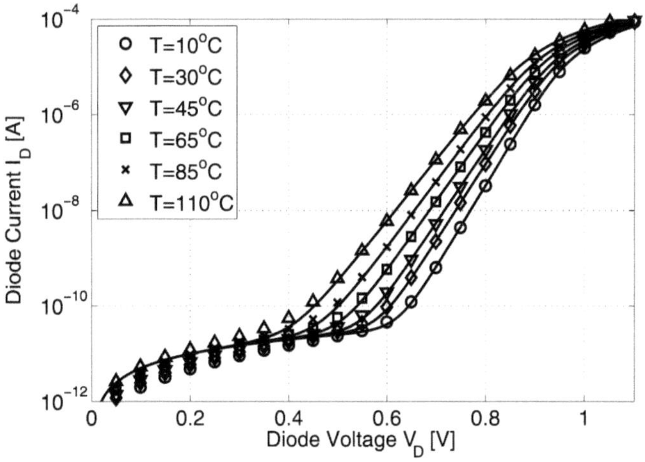

Fig. 4.10 Measurement (*symbols*) and simulation (*lines*) of gated diode $I-V$ characteristics for different temperatures

low current densities to the drift/diffusion region with the typical strong exponential dependence occurs at slightly higher voltages.

For circuit simulation a model has been derived as shown in Fig. 4.9. The exponential dependence of the diode current is provided by a voltage controlled current source I_{Dio} following

$$I_{Dio} = I_S \exp\left(\frac{V_D}{n V_{the}}\right). \tag{4.12}$$

The generation/recombination part of the characteristics is approximated by a leakage resistor R_{Leak} in parallel. As shown later this region does not affect the bandgap performance significantly. At high current levels the series resistance R_S limits the current. For the design of a voltage reference circuit accurate modeling of the temperature behavior is required. In this model the temperature dependence is considered by a variable saturation current $I_S(T)$ [99]:

$$I_S = I_S(T_{nom}) \cdot \left(\frac{T}{T_{nom}}\right)^{\frac{P \cdot T}{n}} \cdot \exp\left(\frac{E_G(T/T_{nom} - 1)}{n \cdot V_{the}}\right) \tag{4.13}$$

with the fitting parameters $n = 1.03$, $P = 2.9$ and the bandgap energy E_G. R_{Leak} is about 10^{11} Ω and assumed to be temperature independent. All model parameters have been extracted from DC and temperature measurements [100]. As shown in Fig. 4.10 measurement and simulation are in good agreement. The model-hardware correlation in the generation/recombination part of the characteristics could be further improved taking the temperature dependence of R_{Leak} into account.

4.3.2 Low Voltage Bandgap Reference

In 1999 Banba et al. proposed a bandgap reference suitable for low-voltage operation as shown in Fig. 4.11 [101]. In this circuit two currents featuring NTC and PTC rather than the corresponding voltages are used for temperature compensation.

The pFETs M_1, M_2 and M_3 have the same dimensions and are controlled by the output of the OpAmp, i.e. the currents I_1, I_2 and I_3 are equal. Moreover, the resistors R_1 and R_2 have the same value and the OpAmp ensures $V_1 = V_2$. Consequently the currents through the resistors R_1 and R_2 are equal. Therefore also the diode currents have to be equal, i.e.

$$I_{1a} = I_{2a} \quad \text{and} \quad I_{1b} = I_{2b}. \tag{4.14}$$

The voltage drop over R_3 represents the voltage difference of the single diode D_1 and the m diodes in parallel. In other words

$$I_{2b} = \sum_{j=1}^{m} I_{D2j} = \frac{\Delta V_D}{R_3} = \frac{V_{the}\ln(m)}{R_3}, \tag{4.15}$$

providing the PTC. The NTC is provided by $I_{2a} = V_D/R_2$. I_2 is the sum of both currents which is mirrored to I_3, hence:

$$V_{ref} = R_4\left(\frac{\Delta V_D}{R_3} + \frac{V_D}{R_2}\right). \tag{4.16}$$

To compensate for the temperature dependence of V_{ref} the ratio of R_2/R_3 has to be chosen properly.

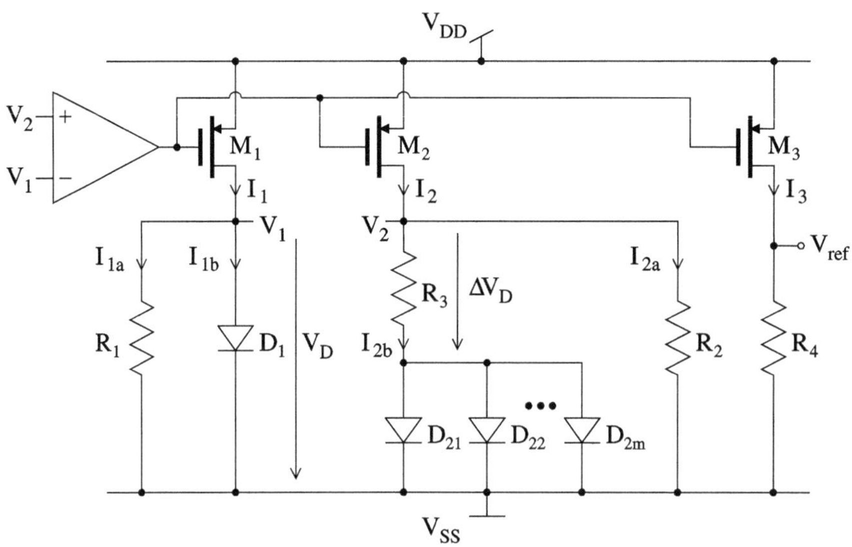

Fig. 4.11 Schematic of low voltage bandgap reference circuit

To ensure appropriate operation of the circuit two conditions have to be fulfilled. The diodes have to stay in the steep exponential part of the I–V characteristics over the whole temperature range and the current source pFETs have to be operated in saturation region. Both conditions limit the downscaling of the supply voltage V_{DD}:

Scaling down V_{DD} and $V_{1/2}$ with the same factor decreases the diode currents exponentially. To keep the currents, i.e. the circuit behavior constant the diode area has to be increased exponentially. Scaling down all currents is no attractive solution neither, since exponentially decreasing currents $I_{1/2}$ combined with linear decreasing voltages $V_{1/2}$ result in exponentially increasing resistor values and areas. Moreover $V_{1/2}$ is limited by the on-set of the generation/recombination part of the I–V characteristics (about 0.8 V at −40°C extrapolated from measurements in Fig. 4.10).

Scaling V_{DD} with constant $V_{1/2}$ lowers the drain-source voltage of the current source pFETs. As soon as they are biased with low overdrive at the limit to the linear region, their output conductance g_{ds} increases significantly, affecting the reference voltage and the temperature coefficient: the circuit is designed to keep the currents and therefore V_{ref} constant over temperature. Increasing temperature consequently yields lower $V_{1/2}$ and higher V_{ds} of M_1 and M_2. Depending on the g_{ds} of the current sources the OpAmp has to increase its output voltage more or less significantly to keep I_1 and I_2 constant. The changing OpAmp output voltage obviously affects the reference voltage. Thus, low g_{ds} and sufficient V_{ds} of the current source transistors is required.

Following the relations explained above the circuit has been optimized for a V_{DD} of 1 V. The operational amplifier is realized as standard, two-stage Miller compensated amplifier as shown in Fig. 4.6. The OpAmp achieves more than 60 dB open loop gain in simulations due to the high g_m/g_{ds} of the FinFETs. The simulation in Fig. 4.12(a) shows that the circuit starts to operate properly at about 0.85 V. Here the feedback loop starts to keep V_{ref} constant. The simulated power supply rejection ratio at nominal V_{DD} is 41 dB. The simulated temperature dependence is shown in Fig. 4.12(b) for different supply voltages. At nominal V_{DD} of 1 V the temperature

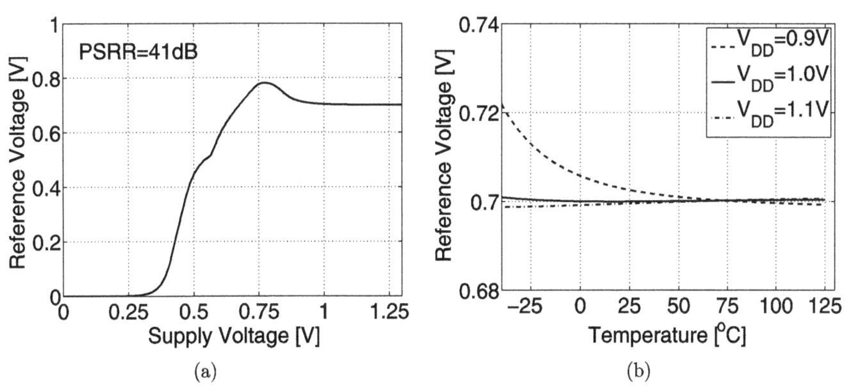

Fig. 4.12 Simulated DC response of V_{ref} on V_{DD} (**a**) and temperature (**b**)

coefficient is smaller than 20 μV/°C. Lowering the nominal V_{DD} of 1 V by 100 mV results in strongly degraded temperature performance due to the degraded g_{ds} of the pFETs as explained above. Increasing V_{DD} does not affect circuit behavior significantly.

4.3.3 Design Considerations

4.3.3.1 Output Conductance of Current Sources

To determine the impact of the finite g_{ds} of the current source pFETs quantitatively, they have been replaced in simulation by an ideal voltage controlled current source in parallel with a resistor. The simulated PSRR is more or less constant over g_{ds} until it starts dropping drastically at a g_{ds} value of about 0.1 μS, see Fig. 4.13(a). Instead of decoupling V_{ref} from V_{DD} the resistive voltage divider of R_4 and M_3 dominates then. The output conductance g_{ds3} is no longer negligible compared to the resistor. The absolute value of the PSRR is higher using a controlled current source with resistor in parallel than using a pFET with equal g_{ds}. The reason is the missing transition from saturation to linear regime in the idealized approximation. The temperature compensation is also affected by finite g_{ds}: the TC starts to degrade for even lower g_{ds} values. Keeping g_{ds} as low as possible is key for sufficient bandgap performance. The use of FinFETs with superior analog properties is beneficial here.

4.3.3.2 Leakage Resistor in p-i-n Diode

Measurements show that the transition of generation/recombination regime to steep exponential behavior of the p-i-n diode occurs at slightly higher voltages compared to standard well p-n diodes. In the model this is reflected by a lower absolute value of

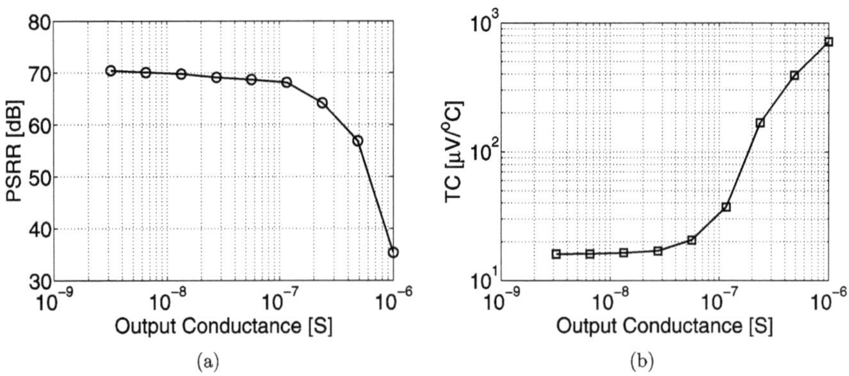

Fig. 4.13 Simulated PSRR (**a**) and TC (**b**) for varying g_{ds} of pFET current sources

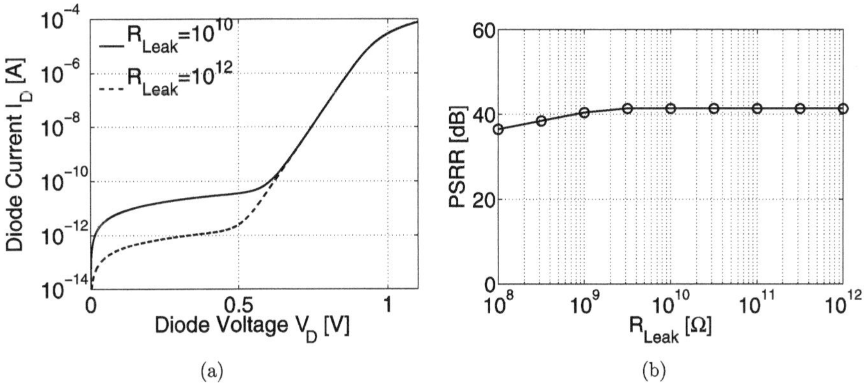

Fig. 4.14 Measured p-i-n diode I–V curves with max. and min. R_{Leak} (**a**) and simulated PSRR for varying R_{Leak} (**b**)

the leakage resistor. Moreover larger variations of R_{Leak} from sample to sample have been observed in measurements. R_{Leak} varies in measurements at room temperature from 10^{10} Ω to 10^{12} Ω, see Fig. 4.14(a), whereas typical values of R_{Leak} for p^+ in n-well diodes are larger than 10^{13} Ω. The impact of the leakage resistor on circuit behavior is shown in Fig. 4.14(b) and Fig. 4.15(a). For $R_{Leak} > 10^9$ Ω the PSRR as well as the TC are not affected. For values below 10^9 Ω, the TC is degraded significantly, because the diodes are leaving the steep exponential regime. However, 10^9 Ω is still one order of magnitude lower than measured for the worst samples. Consequently the increased R_{Leak} is no concern for multi-gate bandgap references.

4.3.3.3 Op-Amp Offset and Gain

Finite open loop gain and offset voltage are considered as main OpAmp non-idealities. A non-zero OpAmp offset is caused by any mismatch of the differential input pair devices and results in a voltage difference of V_1 and V_2, i.e. $V_1 = V_2 + V_{os}$. Taking V_{os} into account the reference voltage changes to:

$$V_{ref} \approx R_4 \left[\frac{\Delta V_D}{R_3} + \frac{V_D}{R_2} - \left(\frac{1}{R_2} + \frac{1}{R_3} \right) V_{os} \right]. \tag{4.17}$$

Depending on the resistor values even an amplified offset V_{os} may be added to V_{ref}. In the bandgap reference considered here, the gain is about 4, i.e. an offset of 1 mV changes V_{ref} by about 4 mV. Thus, low offset voltages are strongly required for sufficient bandgap performance. Moreover the offset voltage affects also the temperature compensation as shown in Fig. 4.15(b). However, compared to the impact on the absolute value of V_{ref}, the TC increases only slightly. The PSRR is almost not affected by V_{os}, simulations reveal a 1 dB degradation for an offset of 5 mV.

Similar to an offset voltage, finite open loop gain results in a non-zero difference of $V_1 - V_2$. To analyze the impact of finite open loop gain the OpAmp has been

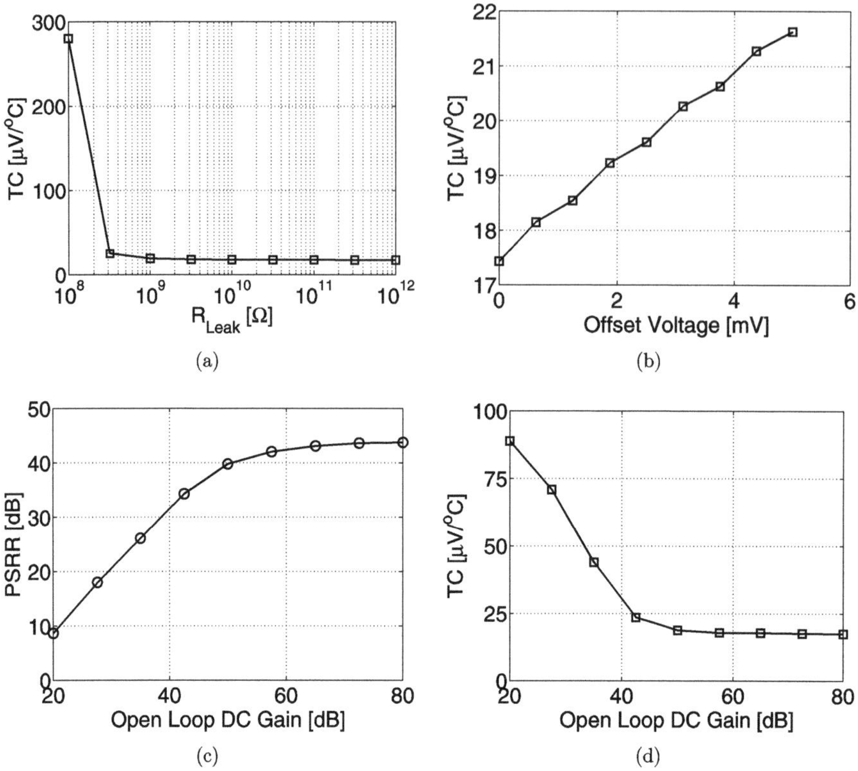

Fig. 4.15 Simulated TC for varying R_{Leak} (**a**), simulated TC for varying offset of the OpAmp (**b**), simulated PSRR (**c**) and TC (**d**) for varying open loop gain of the OpAmp

replaced for simulations with an ideal voltage controlled voltage source. As shown in Fig. 4.15(c), the PSRR is very sensitive to the open loop gain. Increasing the OpAmp gain beyond 60 dB does not improve the PSRR further since the voltage divider of R_4 and M_3 limits the PSRR. The temperature compensation is less affected by finite OpAmp gain. As shown in Fig. 4.15(d), the degradation of TC starts for open loop gain values below 40 dB.

4.3.3.4 Impact of Process Variations

The impact of process variations is investigated by Monte Carlo (MC) simulations. The Monte Carlo model parameters for the FinFET devices (V_T and μC_{ox}), the resistor values and the leakage resistor of the diodes are extracted from measurements [71, 100]. Figures 4.16(a) and (b) shows the resulting distribution of V_{ref} for a supply voltage of 1.0 V and 0.9 V. The reduction of V_{DD} does almost not affect the distribution of V_{ref}, the relative standard deviations are 7.5% and 7.9%, respectively. The main contributors to the statistical distribution are the two input devices

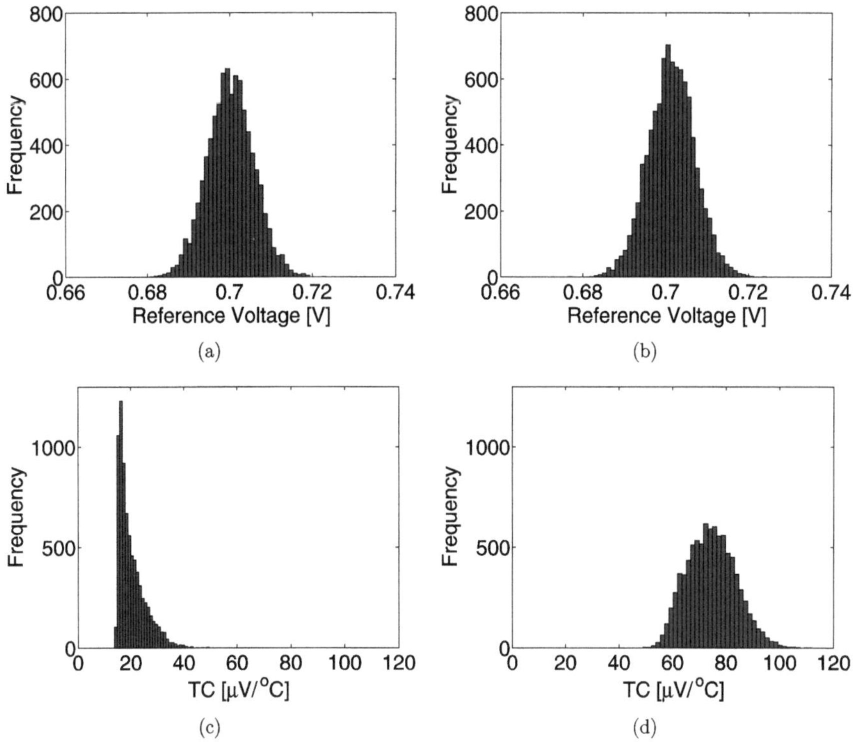

Fig. 4.16 Distribution of V_{ref} for $V_{DD} = 1.0$ V (**a**) and 0.9 V (**b**) and distribution of TC for $V_{DD} = 1.0$ V (**c**) and 0.9 V (**d**)

of the OpAmp (≈20% each), followed by the three current source devices (≈15% each), the resistors (<3% each) and finally the diode leakage resistor (<1% each). Thus, the large spread of the gated p-i-n diode R_{Leak} is no concern, since it does not impact the distribution of V_{ref}.

The simulated distribution of the TC for a V_{DD} of 1.0 V and 0.9 V is shown in Figs. 4.16(c) and (d). As derived above the TC is quite sensitive to the reduction of V_{DD}. On the one hand the absolute value of the TC is increased, one the other hand also the shape of the distribution is broadened. In other words, the reduction of V_{DD} increases the sensitivity of the TC against process variations significantly.

4.3.4 Measurement Results

The low-voltage bandgap has been implemented in a FinFET testchip using gated p-i-n diodes and TaN resistors [102]. The measured circuit performance is shown in Fig. 4.17 and Table 4.3. The DC response of V_{ref} on V_{DD} is close to simulation. The absolute value of V_{ref} at nominal V_{DD} is 0.75 V, a PSRR of 40 dB is

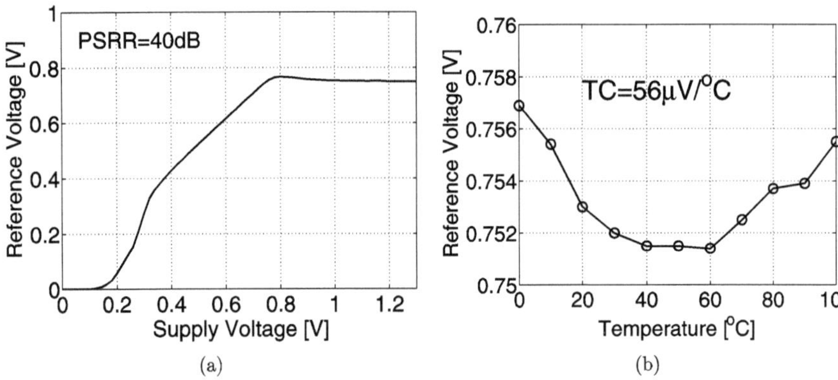

Fig. 4.17 Measured dependence of V_{ref} on V_{DD} (**a**) and temperature (**b**)

Table 4.3 Performance summary of BGR

Characteristic	Value
Supply voltage	1.0 V
Power consumption	100 μW
Reference voltage	0.75 V
Power supply rejection ratio	40 dB
Temperature coefficient	56 μV/°C
Active area	0.035 mm^2
OpAmp DC gain	≈60 dB

obtained. 5 samples of the OpAmp have been characterized stand alone, yielding a mean open loop gain of more than 60 dB. The relation of measured PSRR of the BGR to open loop gain of the OpAmp is in-line with the simulations shown above. The measured temperature coefficient is 56 μV/°C. The difference to simulation results from slightly off-target resistors. Optimization of the TC is easily possible by adjusting the resistor values. It is noteworthy that the measured curvature of the temperature characteristics shows a minimum instead of the maximum which is derived from calculations and observed in most cases [45]. This effect is caused by strong non-linearities of the current source devices. Due to the low supply voltage they need to be operated in a bias point close to the transistion region from linear to saturation.

The circuit area is dominated by the resistors and diodes. The power consumption of 100 μW is determined by the OpAmp. To enable stand alone characterization of the OpAmp, its load driving capability is higher than required by the bandgap reference.

The experimental results prove the feasibility of low-voltage bandgap reference circuits using gated p-i-n diodes in emerging multi-gate CMOS technologies. However, a nominal V_{DD} of about 0.9 V–1 V seems to be the lower limit for the concept shown here. Again, the beneficial FinFET analog and matching behavior improves circuit performance.

4.4 D/A Converter

The feasibility of analog building blocks in FinFET technologies has been proven in the previous sections. Next mixed-signal circuits are discussed. Compared to analog and RF the complexity of mixed-signal circuits typically is much higher. Besides analog and RF building blocks such as OpAmps, comparators, bandgaps or VCOs also digital circuitry is required. Due to the large device count and the broad range of device dimensions the technology requirements in terms of matching are very tight. In this context a digital-to-analog converter (DAC) serves as technology test vehicle for mixed-signal applications. The main goals of the design are to show the feasibility of complex mixed-signal circuits and to verify the beneficial FinFET analog and matching properties on circuit level.

For signal processing applications typically current steering D/A converters are used, based on an array of matched current sources that are switched to the output according to a digital input word, see Fig. 4.18. A segmentation into binary and unary (thermometer) coded part enables a reasonable trade-off in terms of resolution, area and power consumption [103]. The combination of unary and binary coded bits results in a simple and robust design with inherently good accuracy. To prevent mismatch of the binary and unary part, the sum of the binary scaled currents is derived from the unary scaled current sources.

The static performance of the converter is characterized by the differential (DNL) and integral non-linearity (INL). The DNL is defined as the deviation of the actual analog output step from an ideal LSB step given by the smallest binary scaled current. DNL errors are caused by any mismatch of the current sources. The INL is

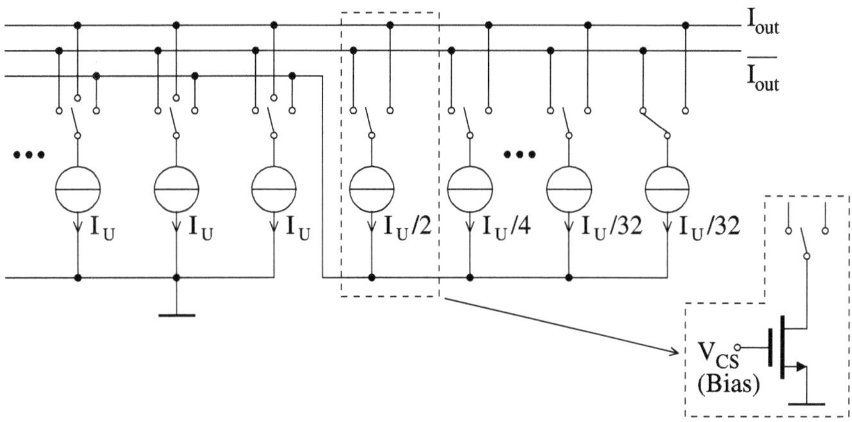

Fig. 4.18 Schematic of segmented D/A converter with exemplary implementation of binary scaled current source

calculated as the deviation of the actual analog output from a linear fit through the real DAC transfer function. Thus, gain and offset errors are not reflected by the INL. The INL contains the integrated DNL errors but also systematic imperfections of the current sources, caused for example by layout dependent gradients or finite current source impedance. Dynamic performance metrics are the spurious-free-dynamic-range (SFDR), signal-to-noise-ratio (SNR) and signal-to-noise-and-distortion ratio (SNDR). These quantities characterize limitations of dynamic DAC performance, e.g. due to harmonic distortion, synchronization and timing errors of the switch control signals or capacitive feedthrough of clock and data signals to the output. Harmonic distortion is caused by a non-linear transfer curve, which is quantified by the INL error. Thus, the INL is a metric to describe both, static and dynamic DAC performance and consequently a major design target.

4.4.1 Design Considerations

Obviously the converter performance is determined by the current sources. Analytical models allow a translation of DAC specifications into device level requirements in terms of matching, output impedance, noise and parasitic capacitance C_{par} at the output of the current source [104–106]. To outline FinFET specifics in current source design, matching and impedance requirements are used as example.

4.4.1.1 Matching

As mentioned above the random variations of current sources contribute significantly to INL error. Since current source mismatch is statistically distributed, the INL degradation can only be predicted within certain confidence boundaries. For this purpose the concept of INL yield is introduced. The INL yield describes the probability to achieve a given INL requirement (usually 0.5 LSB). To achieve a certain yield for an n-bit DAC, the maximum tolerable standard deviation of a unit current source can be calculated using the inverse cumulative normal distribution[3] (inv_norm) [106]:

$$\frac{\sigma(I)}{I} \leq \frac{1}{2C\sqrt{2^n}}; \quad C = \text{inv\_norm}\left(0.5 + \frac{\text{yield}}{2}\right). \tag{4.18}$$

The relation of MOSFET current mismatch and device dimensions as shown in Chap. 1 now enables to determine the minimum current source area to fulfill the matching specifications:

$$(W \cdot L)_{min} = \frac{1}{2(\frac{\sigma(I)}{I})^2}\left(A_\beta^2 + \frac{4A_{VT}^2}{(V_{GS} - V_T)^2}\right). \tag{4.19}$$

[3] Assuming a normal distribution of current sources.

The beneficial matching behavior of FinFETs directly enables a reduction of circuit area, since the overall DAC area is dominated by the current sources. Switches and digital part typically consume less then 20% of the total area, depending on the number of bits and the segmentation. A 10 bit DAC serves as quantitative example, illustrating the benefits: A_β^2 and $V_{GS} - V_T$ are assumed to be equal for FinFET and planar. The digital overhead of both implementations is calculated as 20% of the planar area. Based on Chap. 2 the values for A_{VT} (FinFET) and A_{VT} (Planar) are assumed to be 2.2 mV µm and 3.5 mV µm, respectively. The values result in a 45% reduction of total area for the FinFET implementation.

4.4.1.2 Output Impedance and Parasitic Capacitance

The impedance at the output nodes is given by the parallel connection of the load impedance and a variable number of parallel unit current sources. With increasing output current the output voltage of the current sources decreases due to the rising voltage drop over the load. A finite current source impedance then results in a decrease of the current steps. This effect leads to a flattening of the output current transfer characteristics which reveals the well known quadratic INL curve. To achieve a certain INL specification for an n-bit DAC the minimum current source impedance R_{cs} is given as [91]

$$R_{cs} \geq \frac{I_u R_{Load}^2 n^2}{4\mathrm{INL}}. \tag{4.20}$$

The minimum current source impedance is not only a DC requirement but has to be fulfilled up to the converter bandwidth. Consequently the parasitic capacitance at the output of the current source, i.e. at the tail node of the switch pair has to be minimized. For this reason in many cases cascoded current sources have to be used: the cascode device boosts the output impedance of the current source device by g_m/g_{ds} and decouples the drain of the large area current source device from the critical switch tail node, see Fig. 4.19. Consequently the parasitic capacitance is lowered. Again the advantageous FinFET analog properties facilitate the design: the required output impedance is reached at shorter gate lengths of current source and cascode device, resulting lower parasitic capacitance at the current source output.

4.4.2 Measurement Results

Following the design guidelines a 10 bit FinFET current steering DAC is implemented. The basic structure is the same as shown in Fig. 4.18. The DAC is segmented into 5 binary and 5 unary coded bits. The current sources are implemented as cascoded nFETs, see Fig. 4.19. To minimize glitches and ensure proper timing a switch driver circuit with D-Flip-Flop, differential latch and buffer is required.

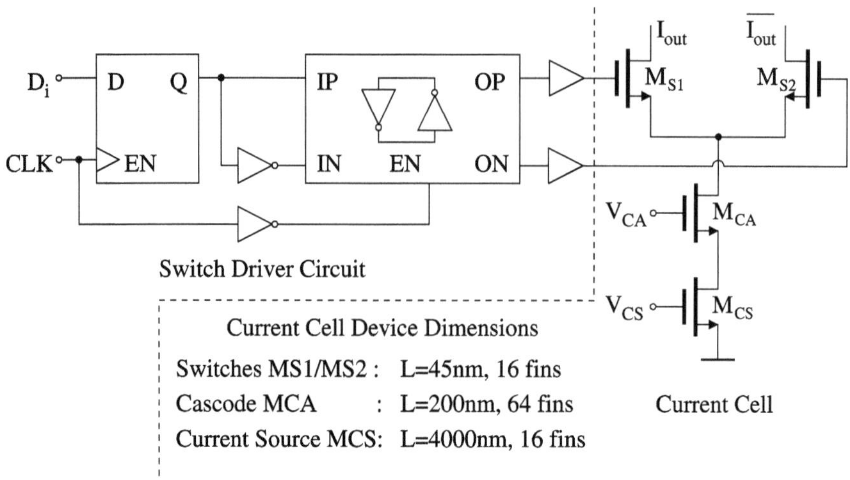

Fig. 4.19 Schematic of current cell and switch driver circuit

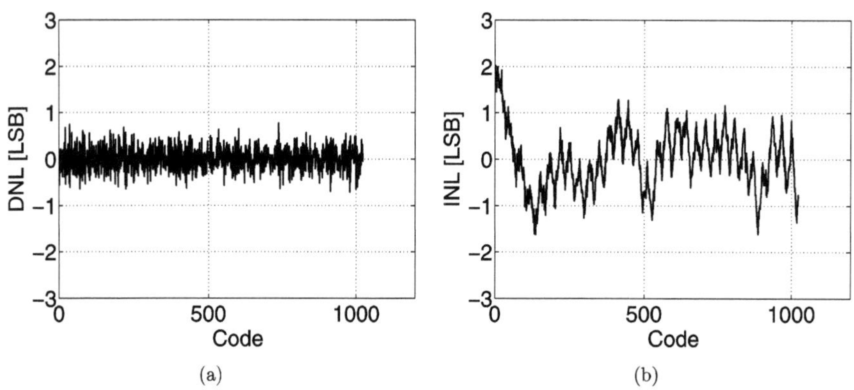

Fig. 4.20 Measured differential- (**a**) and integral non-linearity (**b**), $V_{DD} = 1.0$ V

Taking advantage of the beneficial analog and matching properties the circuit is optimized for minimum area. The layout of the current sources is symmetrical and based on the separated unit cell approach as discussed in Sect. 4.1.1. The overall active area is only 0.018 mm^2, which is significantly smaller than reported for comparable planar designs [107].

Static performance measurements are shown in Fig. 4.20. At a supply voltage of 1 V the DNL is smaller than 1 LSB, the INL does not exceed 2 LSB. The INL is dominated by mismatch, no quadratic shape due to impedance problems is observed.

A dynamic measurement with digital 1.528 MHz sine wave input and a sample rate (=clock frequency) of 20 MS/s is depicted in Fig. 4.21. Due to the differential structure the even harmonics are very small. The third harmonic limits the SFDR to 56 dB. The contribution of other harmonics to the total harmonic distortion (THD)

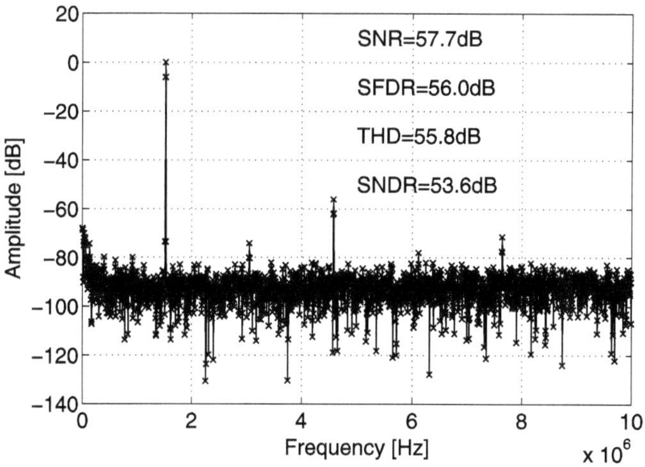

Fig. 4.21 Measured output spectrum for 1.528 MHz signal frequency at 20 MS/s, $V_{DD} = 1.0$ V

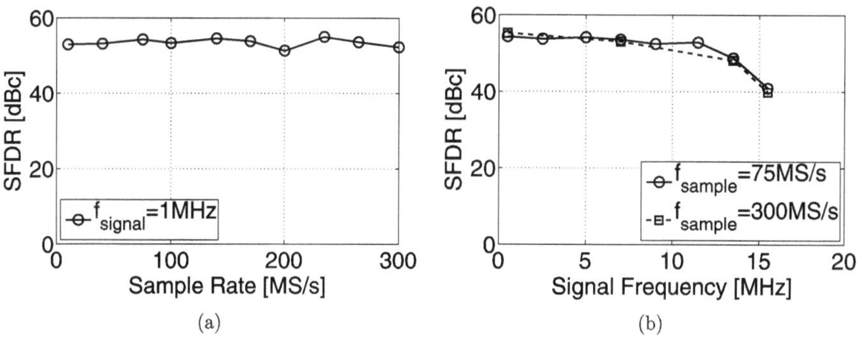

Fig. 4.22 Measured dependence of SFDR on clock frequency (**a**) and signal frequency (**b**)

of 55.8 dB is almost negligible. The measured SNDR of 53.6 dB corresponds to an effective number of bits (ENOB) of 8.6. No significant degradation of SNR due to thermal noise is observed. The measured performance is comparable to planar designs [107]. At 100 MS/s and ±100 μA differential output current the converter draws 380 μA from a 1 V supply.

To find the dynamic limitations of the converter, the SFDR is measured against sample rate and signal frequency. With constant signal frequency no degradation of SFDR is observed until 300 MS/s, see Fig. 4.22(a). However, increasing the signal frequency above 12 MHz leads to a strong reduction of SFDR. As shown in Fig. 4.22(b), the bandwidth limitation is independent of sample rate and most probably caused by underestimated parasitics. Since the active current cell devices are rather small, the SFDR degradation is attributed to underestimated wiring parasitics in layout.

Two main statements are derived from the considerations on FinFET D/A converters: the maturity of recent FinFET technology allows to realize even complex mixed-signal circuits (>1500 devices) and the advantageous analog and matching properties enable reduced circuit area compared to planar designs.

4.5 Phase-Locked-Loop Circuit

Phase-Locked-Loop circuits are widely used in modern communication systems, e.g. as frequency synthesizer for clock generation or regeneration [80]. Figure 4.23 shows a typical "analog" charge pump based PLL, consisting of phase/frequency detector (PFD), charge pump (CP), loop filter (LF), voltage controlled oscillator (VCO) and frequency divider (FD). Using negative feedback the phase of the VCO is adjusted to track the phase of the reference clock f_{ref}, so that the skew between the two PFD inputs is constant. In steady state when the PLL is locked the desired synchronization $f_{out} = N f_{ref}$ is achieved. Besides target frequency range (f_{ref} & f_{out}) and locking time, the PLL performance is specified by power consumption, circuit area and jitter. Jitter describes the timing uncertainties of the clock edges and corresponds to the integrated phase noise of the carrier f_{out} [108].

PLLs contain digital (PFD & FD), analog (CP & LF) and RF (VCO) building blocks. Therefore a charge-pump PLL as shown in Fig. 4.23 is considered as further testvehicle for complex mixed-signal circuits with focus on matching and noise.

4.5.1 Design Considerations

Since there is extensive literature on design optimization for low jitter, e.g. [108–112], here only some fundamental, multi-gate related aspects are discussed.

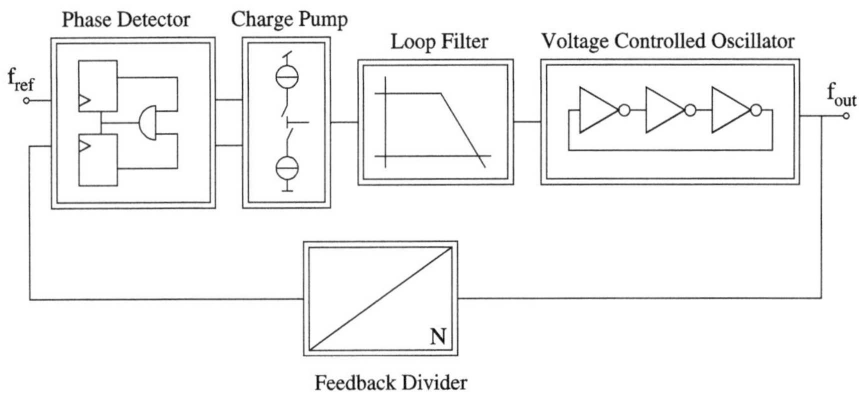

Fig. 4.23 Schematic of analog charge-pump PLL

In principle all blocks contribute to the overall PLL jitter, however reference clock, charge pump, loop filter and VCO typically dominate [113]. Due to the feedback system each block has a different noise transfer function (NTF), i.e. all noise contributions are shaped in a different way. Modeling the PLL as linear continuous time system allows a straightforward calculation of the single noise transfer functions [114, 115]. The main design parameter influencing the NTFs is the PLL bandwidth, which is defined by loop filter, charge pump current, VCO gain (i.e. tuning sensitivity) and divider ratio.

4.5.1.1 Charge Pump

Regarding f_{out} the charge pump noise is low pass filtered, i.e. frequency components below the PLL bandwidth strongly affect overall phase noise and jitter performance. Although charge pump noise is low pass filtered thermal noise typically dominates: the CP output is switched with f_{ref} leading to aliasing of broadband thermal noise, whereas flicker noise is oversampled [111]. However, excessive flicker noise with a corner frequency in the range of f_{ref} would degrade the phase noise performance significantly. Besides noise another important CP design criterion is the mismatch of up and down currents, resulting in spurious tones in the output spectrum. Device parameter mismatch and low output resistance are the main sources for up and down current mismatch. The high FinFET output resistance and the good matching behavior allows to use a simple buffered charge pump without cascodes or regulation as shown in Fig. 4.24.

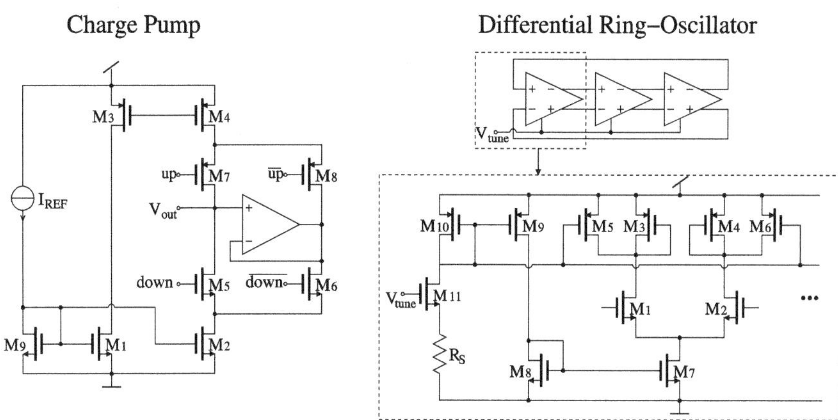

Fig. 4.24 Schematic of single ended, buffered charge pump (*left*) and three stage differential ring oscillator (*right*)

4.5.1.2 Loop Filter

The noise of the loop filter is band pass filtered with a peak at the PLL bandwidth. To simplify matters and facilitate phase noise analysis a passive second order RC filter is used. Thus, accuracy, parasitics and noise are determined by the passive devices. Only thermal noise is generated by the resistors.

4.5.1.3 VCO

The transfer function of the VCO phase noise corresponds to a high pass, i.e. low frequency noise components close to carrier are suppressed. As mentioned earlier the VCO phase noise is determined by up-converted flicker and thermal noise. Potential FinFET related noise issues, such as excessive flicker noise with corner frequency close to PLL bandwidth or increased thermal noise due to high parasitic resistances would be transferred directly to the output and degrade jitter performance. To optimize area and power consumption a ring-oscillator is used instead of a LC-VCO. The oscillator consists of three differential stages and is tuned via the bias current of the tail current sources. A parallel connection of tunable active loads and diode connected loads enlarges tuning range, see Fig. 4.24. Compared to LC-VCOs and single ended implementations the phase noise of differential ring-oscillators is significantly worse[4] [116]. Thus, potential FinFET noise issues would degrade the PLL phase noise even more.

4.5.2 Measurement Results

The charge pump PLL is implemented on a FinFET testchip following the considerations made above. The design is optimized for low power and low active area. The PLL bandwidth is set to 10 MHz as compromise of VCO and CP noise suppression. A fix divider ratio of 8 and a digital three-state phase-frequency detector is used. The target input frequency range is 70 MHz–200 MHz, corresponding to an output frequency of 0.56 GHz–1.6 GHz. To drive the measurement equipment 50 Ω output buffers are integrated. The overall active area including output buffers is 0.043 mm^2, which is dominated by the passive devices (TaN resistors and MIM capacitors) of the loop filter.

The measured phase noise profile at $f_{ref} = 100$ MHz and the rms accumulated jitter is shown in Fig. 4.25. At 1 MHz offset frequency a phase noise of -112 dBc/Hz is achieved. The corresponding rms long-term jitter is 12.63 ps, periodic and integrated jitter are 1.69 ps and 8.93 ps respectively. Beyond the PLL bandwidth the noise suppression is limited to about -140 dBc by the noise floor of the spectrum analyzer. The measured performance is summarized in Table 4.4.

[4]Of course the differential structure has other advantages like improved PSRR, which are however not FinFET related.

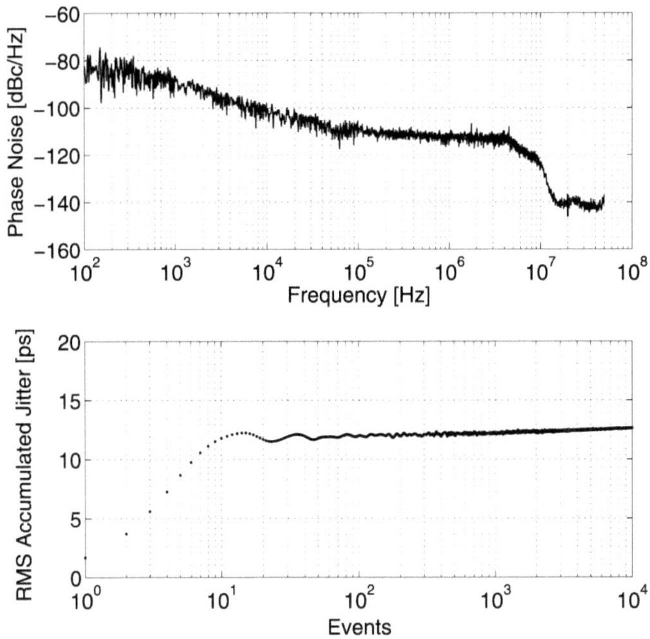

Fig. 4.25 Measured phase noise and rms accumulated jitter at $V_{DD} = 1.0$ V

Table 4.4 Summary of measured PLL performance

Characteristic	Value	Condition
Supply voltage	1.0 V	
Power consumption	3 mW	$f_{ref} = 100$ MHz
Power consumption (incl. buffer)	15 mW	$f_{ref} = 100$ MHz
Capture range	65–225 MHz	
Phase noise	−112 dBc/Hz	@ 1 MHz Offset, $f_{ref} = 100$ MHz
Periodic jitter (RMS)	1.69 ps	$f_{ref} = 100$ MHz
Long-term jitter (RMS)	12.63 ps	$f_{ref} = 100$ MHz
Integrated jitter (RMS)	8.93 ps	$f_{ref} = 100$ MHz
Active area	0.043 mm^2	

For a fair assessment of the achieved PLL jitter and phase noise performance, the power and area consumption have to be taken into account. Compared to planar state-of-the-art PLL designs with similar frequency range as summarized in [112] or [110] the measured figure-of-merit, defined as product of power consumption and integrated jitter is competitive, see Fig. 4.26. No degradation of jitter due to high FinFET flicker or thermal noise is observed.

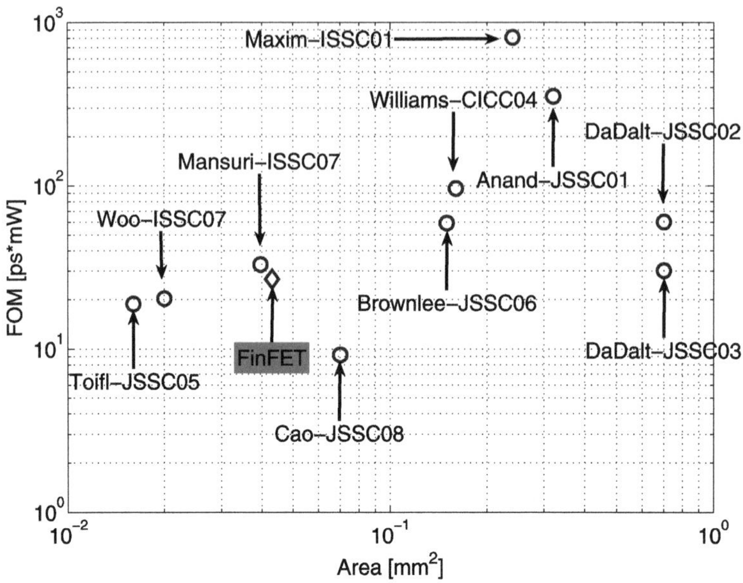

Fig. 4.26 Comparison of FinFET PLL performance to state-of-the-art planar designs [110, 112]

4.6 RF Building Blocks

The ability to realize RF circuits with reasonable performance and power consumption is of outmost importance for the integration of system on chip applications. To give an outlook to FinFET RF design issues, the implementation of VCO and LNA as important building blocks is briefly discussed for a 2 GHz application [117, 118]. The main device related concerns are reduced speed (f_t, f_{max}) due to high parasitics and increased noise (flicker, thermal and gate noise) due to the fin roughness and the interface quality in the high-k, metal gate stack.

4.6.1 LC-VCO

Especially in wireless communication systems VCOs are extensively used in transmit and receive path [79]. In case of tight phase noise requirements LC tank based VCOs are preferred [119]. Important design criteria are center frequency, tuning range, phase noise and power consumption. An LC-VCO with n-type FinFET cross coupled pair and biasing current source is implemented as test vehicle, as illustrated in Fig. 4.27. Frequency tuning is performed with varactors consisting also of n-type FinFETs. To drive the measurement equipment 50 Ω pad buffers are integrated. All devices have a gate length of 70 nm and the minimum fin width of 30 nm. To improve matching the devices are separated in unit cells of 6 fins, which are arranged

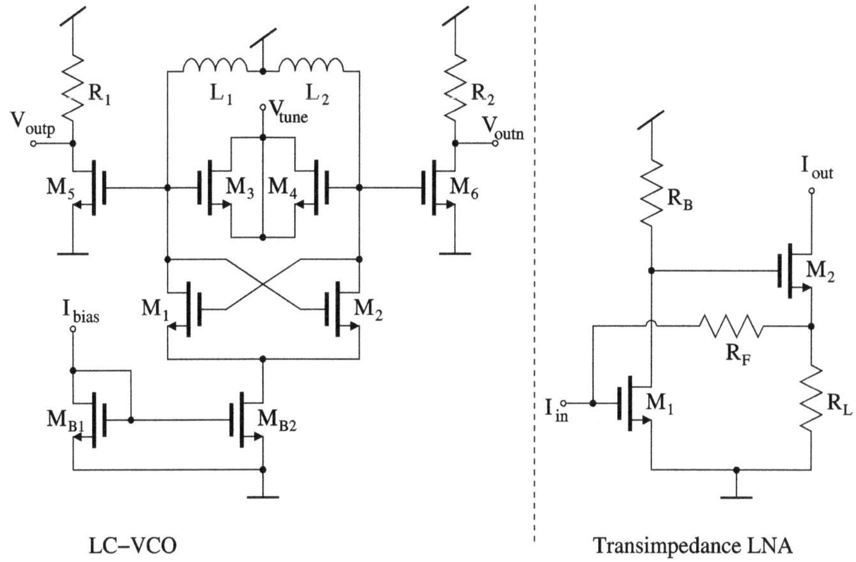

Fig. 4.27 Schematic of LC-VCO (*left*) and feedback based transimpedance LNA (*right*)

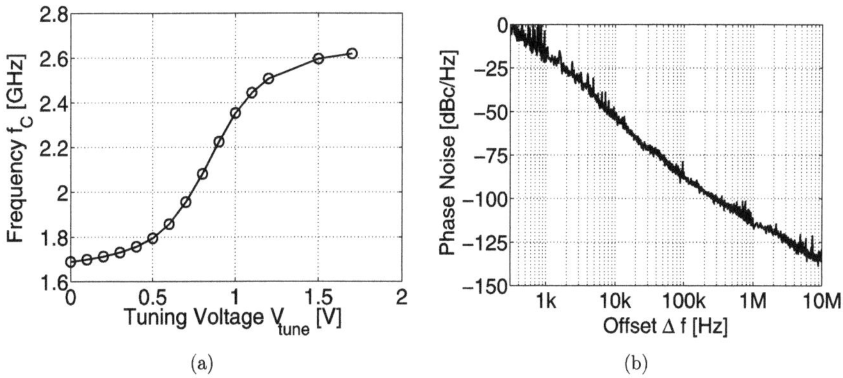

Fig. 4.28 Measured tuning range (**a**) and phase noise spectrum at f_c (**b**), $V_{DD} = 1.2$ V

in a dense array [117]. Due to the BEOL restrictions in the testchip the differential inductor is build with 3 thin metal layers, limiting the quality factor Q. 3D EM simulations reveal a differential Q of about 5.

All measurements are done with a supply voltage of 1.2 V. The VCO consumes 12 mA including the pad buffers. The center frequency f_c of 2.2 GHz is close to target of 2 GHz. As shown in Fig. 4.28(a) a tuning range of 0.8 GHz is achieved, corresponding to 36% of f_c. The tuning range is sufficient for typical GSM specifications. The measured phase noise spectrum at f_c is shown in Fig. 4.28(b). At an offset frequency of 3 MHz the phase noise is -123 dBc/Hz, which is not competitive to similar planar designs [119]. A small fraction of the high phase noise level is attributed to an increased flicker noise level resulting from the not-optimized high-k gate stack available for this testchip: the transition from flicker noise limited region with f^{-3} dependence to thermal noise region with f^{-2} dependence is observed at about 200 kHz–300 kHz, which is higher than reported for the planar reference realized in 0.25 μm CMOS with SiO_2 dielectric. However the high phase noise is mainly attributed to the low quality factor of the inductor. Another reason for phase noise degradation is the minimum fin width of the varactors, yielding a high voltage sensitivity $\partial C/\partial V$ [120]. Thus, competitive phase noise performance could be achievable with RF capable BEOL and optimized varactors.

4.6.2 LNA

Low noise amplifiers are required as interface of the receive path to the antenna or the duplexer [109]. They are intended to generate a sufficient input signal amplitude for the following blocks like mixer or filter without adding too much noise. Hence, important performance metrics of LNAs besides linearity, area and power consumption are gain, noise figure and input matching. To avoid the need for inductors a broadband LNA based on a feedback transimpedance topology is implemented, as shown in Fig. 4.27 [121]. The n-type FinFETs feature a gate length of 70 nm and a fin width of 30 nm. The main drawback of the topology is the high power consumption needed to provide proper input matching [122]. To reach the high W/L values for the input matching, again large arrays of 6 fin unit cells are used. At 2 GHz the LNA draws 25 mA from a 1.2 V supply. It is noteworthy that this high power dissipation could cause considerable self-heating effects in the dense device arrays, as discussed in the next section. Thus, layout optimization may reduce self-heating and power consumption.

The measured S-parameters are shown in Fig. 4.29(a). At 2 GHz a gain (S_{21}) of about 15 dB is achieved, the corresponding input matching (S_{11}) is better then

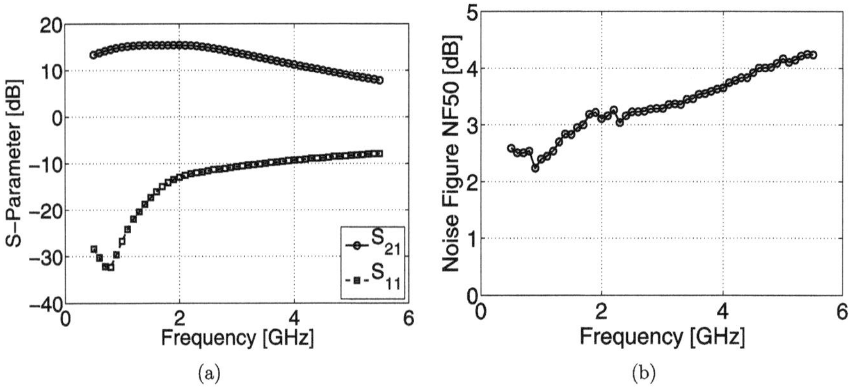

(a) (b)

Fig. 4.29 Measured S-parameter (**a**) and noise figure at 50 Ω versus frequency (**b**), $V_{DD} = 1.2$ V

-12 dB. The broadband nature allows operation until 4 GHz with a gain of more than 10 dB and input matching around -10 dB. The noise figure at 50 Ω (NF50) is shown in Fig. 4.29(b). At the target frequency of 2 GHz the noise figure is 3.1 dB, at 4 GHz still a NF50 of 3.6 dB is achieved. Linearity and compression point are not measured due to the limitations of the on-wafer measurement setup. The measured performance is sufficient for UMTS applications. For a fair comparison a figure-of-merit comprising gain, noise figure and power consumption is introduced:

$$\mathrm{FOM} = \frac{\mathrm{S}_{21}}{\mathrm{NF50} \cdot \mathrm{Power}}. \tag{4.21}$$

Compared to a planar bulk implementation reported in [122], which is based on the same topology, no performance degradation is observed. The figure-of-merit is even improved from 0.135 mW^{-1} to 0.161 mW^{-1}.

Summarizing the results on RF building blocks, no real show stopper for RF applications in the sub 10 GHz regime was identified, confirming statements in [123] and [124].

4.7 Self-Heating

The impact of self-heating on circuit design is discussed next. As shown in Chap. 2 power dissipation in active devices results in local temperature increase which affects the actual transistor behavior. Different aspects of self heating have to be considered:

- *Self-heating of single device*: in general every device suffers from self-heating if it is biased with certain power density. Consequently the device behavior is affected also from DC perspective, depending on the bias conditions. At high power density the current is reduced due to the lowered carrier mobility. From circuit perspective this effect needs to be compensated by higher bias currents to maintain specified device parameters such as g_m, resulting in higher overall power consumption. If a device is modulated from bias point also the self-heating changes. This can lead for example to an increase in output conductance [50]. However, using dedicated models as presented in Chap. 2 the circuit design can be adapted accordingly.
- *Thermal coupling*: strong heating of single devices can impact the local device temperature of other neighboring devices. Depending on the purpose of the device this effect may be desired or not: e.g. changing the temperature of a single device within a current mirror causes a mismatch of output currents. If one transistor of a differential pair heats up the other one matching is actually improved.
- *Transient thermal mismatch*: similar to charge trapping self-heating causes a dependence of device behavior on operation history. Devices with different operation history therefore show transient mismatch, e.g. in a differential input pair.

4.7.1 Thermal Coupling

Layout dependent thermal coupling has already been reported for planar SOI devices [125]. To assess the effect for multi-gate SOI devices a test-structure is developed, as illustrated in Fig. 4.30. The n-type FinFET serves as heating device. Drain and gate bias of the FinFET are adjustable off-chip to vary the heating power. The diode is used as temperature sensor and is implemented as gated p-i-n junction similar to the devices shown in Sect. 4.3.1. The diode is reverse biased and the leakage current is measured with a parameter analyzer. The temperature dependence of the leakage current is used to sense temperature changes resulting from changes in heating power. For calibration the temperature characteristics of a reference p-i-n diode is measured first.

Different layout scenarios are compared as illustrated in Fig. 4.30:

- *Shared source/drain areas*: the worst case scenario in terms of thermal coupling is a distance of zero between diode and FinFET, i.e. the source and cathode area are shared.
- *Separated source/drain areas*: the distance between separated unit devices is varied to assess the range of thermal coupling.
- *Separated source/drain areas with dummy heat sinks*: to avoid cross-heating a dummy heat sink is added, consisting of large contact landing pads with dummy contact array, dummy metal and via stack.

The corresponding measurement results are shown in Fig. 4.31. As expected the thermal coupling of the shared source/drain structure is very pronounced, the temperature rises by more than 10°C. A separation of the devices reduces the cross-heating effect, however at 1 μm distance still a significant temperature increase is observed. The use of dummy heat sinks efficiently minimizes thermal-coupling. Compared to the worst case layout cross-heating is reduced by about a factor of ten at 1 mW heating power. Obviously the dummy structure implies an area overhead.

In most cases the consequences on circuit level are not very drastic. In typical analog bias conditions, e.g. $V_{GS} = V_T + 200$ mV, $V_{DS} = V_{DD}/2$, the power density is rather low compared to the measured examples. The corresponding temperature

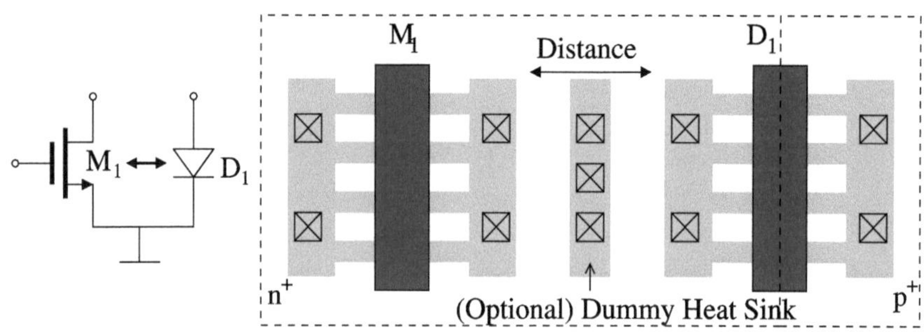

Fig. 4.30 Schematic and layout implementation of cross-heating teststructure

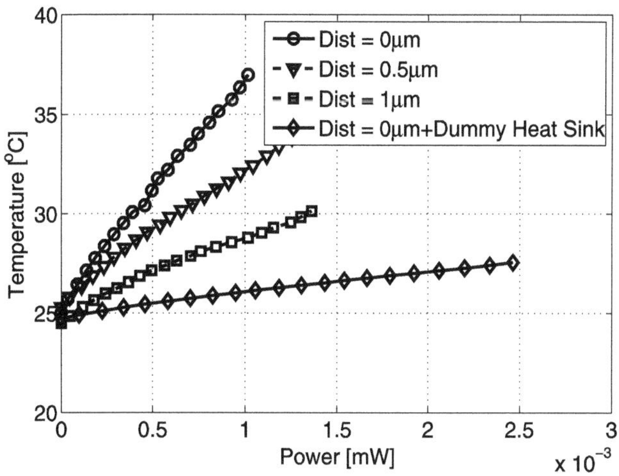

Fig. 4.31 Measured temperature increase due to cross-heating for different distances and layout styles

changes do not exceed few °C. Moreover, MOSFETs biased with a gate source voltage in the range of V_T show a low sensitivity against temperature changes, since V_T reduction and mobility degradation partly compensate each other [126]. Thus, the device mismatch induced by thermal coupling in low-power analog circuits is expected to be of minor importance. In cases with remarkable impact of thermal mismatch, e.g. in dense arrays with high power dissipation like in a LNA, the mismatch can be minimized by careful layout optimization.

4.7.2 Transient Thermal Mismatch

Different bias conditions and operation history of FinFETs yield transient thermal and electrical mismatch. Generally speaking self heating induced transient mismatch impacts analog and mixed-signal circuit behavior in a similar way as hysteresis effects discussed in Chap. 3. Therefore only some fundamental considerations are presented here. Three main differences can be identified comparing self-heating to charge trapping: heating and cooling of devices typically occur with symmetric time constants, i.e. no "charge-pump" effects take place. Moreover the absolute values of the time constants are typically in the range of 10 ns to 100 ns which is significantly shorter than in case of charge trapping. Finally, the transient mismatch is proportional to the power dissipation $I_D \cdot V_{DS}$, not to the gate source voltage V_{GS}.

4.7.3 Linear and Continuous Time Circuits

The impact of self heating induced transient mismatch on linear analog circuits like operational amplifiers is marginal in most cases. The typical analog bias conditions imply moderate to low power density as mentioned above. Moreover the differential input needs to be small for reasonable operation. Both aspects yield very low transient mismatch at the OpAmp input. Large signal swings at the output stage of a differential amplifier may reduce the output amplitude, however harmonic distortion is not affected [127]. Other examples, like slightly increased small signal amplifier gain due to self-heating are given in [75].

4.7.4 Non-Linear and Discrete Time Circuits

Non-linear circuits are more sensitive to self-heating. Again comparators are operated in worst case conditions with respect to self-heating induced transient mismatch. As explained in Chap. 3 incorrect comparator decisions can be caused by switching the differential input from a large to a small value with constant sign [50]. Interestingly this effect depends not only on circuit parameters like gain or sampling frequency but also on the actual comparator architecture, since mismatch is proportional to $I_D \cdot V_{DS}$, not to V_{GS}. Switching from high to low input value in a comparator consisting of differential stages with resistive load of course leads to a significant change of current levels in the input devices. However, the actual drain source voltage of the input devices changes in the opposite way: higher currents increase the voltage drop over the resistor load and reduce V_{DS}, i.e. the change in power dissipation is partly compensated. The situation is different in topologies as shown in Fig. 4.32. Here a two stage comparator is depicted with low-ohmic output nodes in the first stage. The first stage realizes a current difference which is transformed into a voltage difference in the second stage. In this topology the power

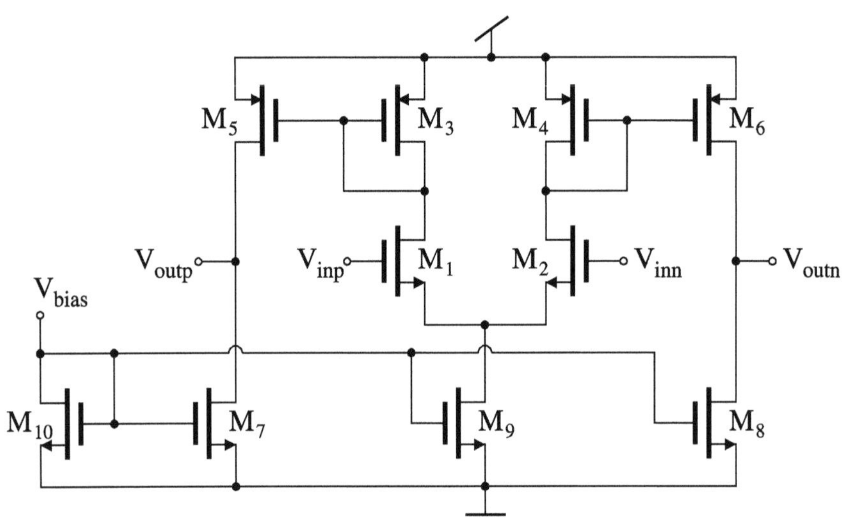

Fig. 4.32 Schematic of two stage comparator with diode connected load in input stage

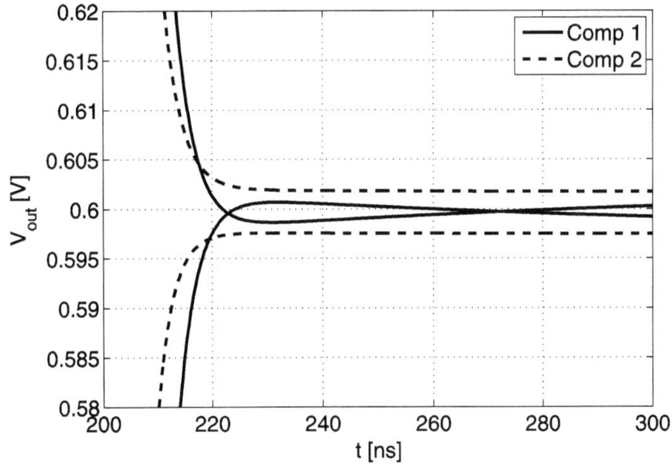

Fig. 4.33 Simulated transient response of current mirror based comparator (comp1) and resistively loaded comparator (comp2)

dissipation of the input devices changes considerably switching from large to small input values, i.e. the circuit is sensitive to self-heating induced transient mismatch.

The difference is illustrated in Fig. 4.33. The transient response of two comparators is compared switching the differential input from large to small value with constant sign. The circuit model introduced in Chap. 2 is used for simulation. Comparator 1 is based on the schematic shown above, whereas the three stage implementation with resistive load shown in Fig. 3.6 is called comparator 2. Obviously the resistively loaded comparator is almost insensitive to self heating, whereas comparator 1 delivers an incorrect decision for about 50 ns until the thermal equilibrium is recovered. Ensuring an efficient thermal coupling in layout reduces the probability for incorrect comparator decisions.

In summary, self-heating is another source of device parameter variation which has to be considered in some cases, especially if high resolution is required. The impact of self heating will rise in future technologies with decreasing active silicon volume and increasing power density, see Chap. 2. Careful layout optimization and choice of topology are suitable measures to reduce the impact of transient mismatch induced by self-heating.

4.8 Selective Fin Width Tuning

The fin width is a crucial device parameter determining the electrical device behavior for two reasons, see Chaps. 1 and 2: high parasitic series resistances in narrow fins limit drive current, g_m and f_t. In addition the gate control over the channel, i.e. the short-channel behavior is determined by the relation of fin width to gate length. The need to minimize leakage currents enforces the use of the minimum fin width in digital circuits. However, from analog and RF perspective there is no optimum fin width [128]: wider fins enable higher g_m and f_t, whereas the worsened channel-control means reduced efficiency g_m/I_D and voltage gain g_m/g_{ds}. Moreover flicker noise depends on fin width. In other words, the fin width represents an additional degree of freedom in analog circuit design. Although lithography restrictions impose a certain granularity, selective tuning of fin width enables novel approaches for circuit optimization.

Two examples are briefly discussed next, based on the fact that introducing short-channel effects by widening fins changes the effective threshold voltage in saturation [128].

4.8.1 Self Cascode

Cascoding is a widely used technique to boost the output impedance and to reduce the Miller effect of a common source transistor, see Fig. 4.34 [39]. To avoid the need for generating an additional bias voltage for the common gate transistor a stacked current source (self cascode) has been proposed [129]. Using the multi V_T option commonly available in recent CMOS technologies a stack comprising a common source regular V_T device and a stacked low V_T device is formed. The drain source voltage of M_1 is then given by the difference of the threshold voltages of M_1 and M_2, typically in the range of 100 mV to 200 mV. If V_{in} is close to V_{T1}, M_1 is still in saturation region and M_2 acts as cascode device. This technique is typically used to boost the output impedance of current sources. Instead of using multiple V_T devices which require additional process steps this effect can be achieved by using devices with varying fin width. Changing the fin width, the V_T can be simply tuned by more than 100 mV if the gate length is below 100 nm, see Fig. 4.35(a). For larger gate lengths the effect is less pronounced, due to the improved intrinsic gate control of the long channel devices.

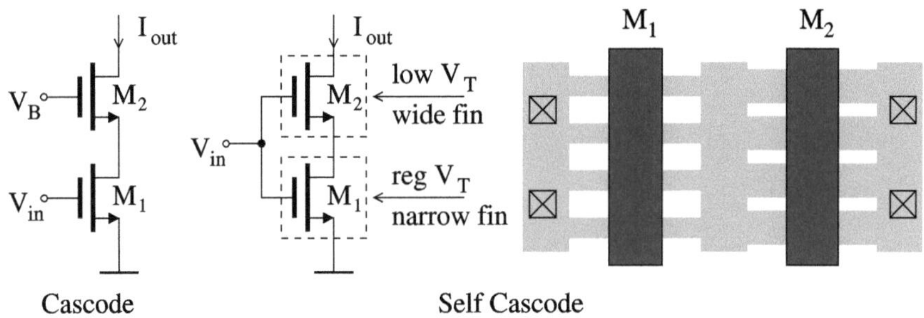

Fig. 4.34 Schematic and layout implementation of stacked current source

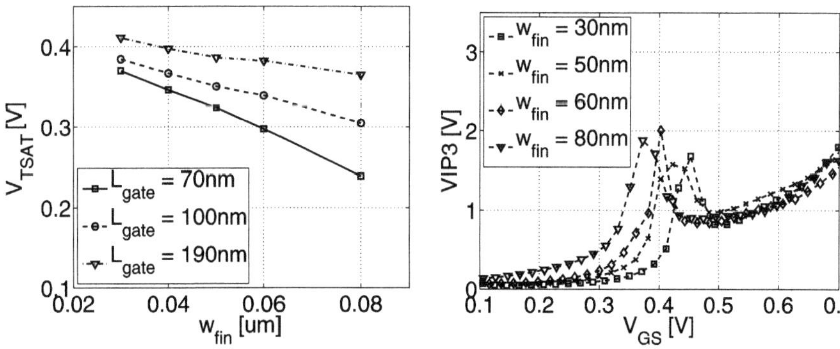

(a) (b)

Fig. 4.35 Measured V_T versus fin width for different L_{gate} (**a**) and measured VIP3 versus V_{GS} for different fin widths and constant $L_{gate} = 70$ nm (**b**)

4.8.2 VIP3 Enhancement

Using multiple fin widths also allows to increase the third order interception point VIP3 according to (1.6). Typical VIP3 characteristics for varying fin width are shown in Fig. 4.35(b). A smooth increase of VIP3 is followed by a sharp local maximum close to V_T (zero crossing of g_{m3}), then VIP3 rises again with V_{GS}. To avoid the uncertainty around the maximum, usually a bias point in the flat region next to the maximum is chosen if linearity is a concern. Since the VIP3 values obtained there are less than the nominal V_{DD} of 1 V, significant non-linearity could occur at maximum output signal swing. However, a combination of different fin widths yields a much broader maximum, since the maximum is shifted with V_T, see Fig. 4.35(b). Thus, biasing a transistor in the broadened VIP3 maximum is no longer affected by small device parameter variations. Obviously the overall device area has to be increased to maintain the same bias current, as the gate source voltage moves closer to V_T.

Going beyond the FinFET specific examples, it has been shown that systematically employing the multi V_T option enables significant benefits for analog circuit design in low supply voltage scenarios [130].

Chapter 5
Multi-Gate Tunneling FETs

To give an outlook to analog design aspects beyond multi-gate CMOS in this chapter implementation and application aspects of multi gate tunneling FETs (MuGTFETs) are discussed. Tunneling FETs are considered as possible device concept for post CMOS technologies, circumventing classical scaling limitations by a novel principle of operation, see Chap. 1. Tunneling FETs are implemented in a state-of-the-art FinFET technology. Basic digital and analog device performance is analyzed. The scaling potential of multi-gate tunneling FETs is proven by measurements and device simulations that reveal a low dependence of the device characteristics on the channel length. Temperature dependence and matching behavior is analyzed in order to assess device variability. A new voltage reference circuit is proposed as potential application for the MuGTFET.

5.1 Principle of Operation and Implementation of MuGTFETs

The TFET structure and principle of operation is illustrated in Fig. 5.1 for an n-channel device in SOI technology. To obtain a tunneling junction, complementary source/drain implants are required. The resulting device structure is equivalent to a gated p-i-n diode. The p-i-n diode is operated in reverse mode. Thus, the drain-source voltage V_{DS} has to be positive or zero to ensure meaningful transistor operation. If no gate-source voltage V_{GS} is applied, the large pn barrier prevents current flow between source and drain. Due to this large barrier very low off-currents are possible compared to MOSFETs [36]. With positive V_{GS} larger than V_T a electron channel is induced, forming a tunneling junction between the p^+-source and the n-channel due to the strong bending of energy bands. Applying a positive V_{DS} results in Zener tunneling and current flow from source to drain. The junction where the tunneling occurs, i.e. where free charge carriers are generated is defined as source. P-type transistor operation is possible in the same structure forming a tunneling junction between the n^+-junction and a p-channel induced by a negative V_{GS}. The gated p-i-n structure implies a short between source and substrate for planar bulk TFETs. In circuits the integrated substrate contact prevents the use of stacked TFETs

M. Fulde, *Variation Aware Analog and Mixed-Signal Circuit Design in Emerging Multi-Gate CMOS Technologies*, Springer Series in Advanced Microelectronics 28,
DOI 10.1007/978-90-481-3280-5_5,

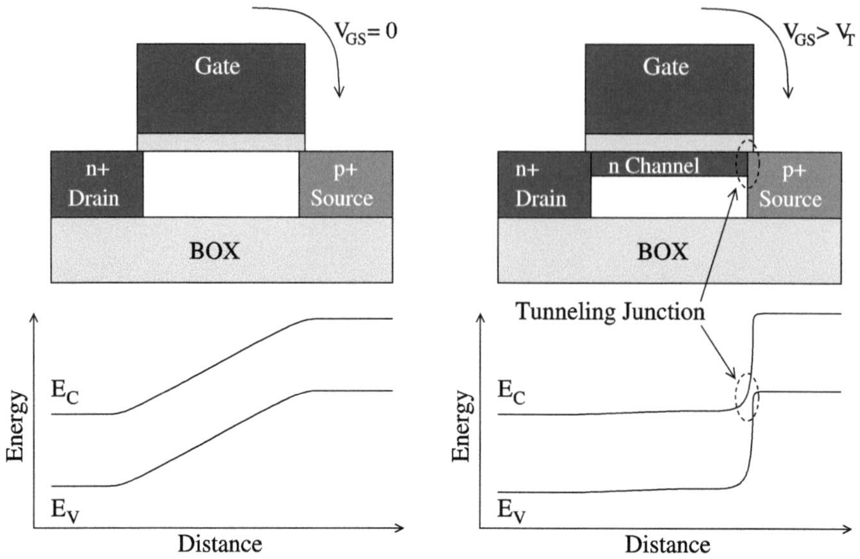

Fig. 5.1 Schematic view and band diagram of n-channel TFET in off- (*left*) and on-state (*right*). Tunneling junction indicated by ellipse

e.g. in logic gates [131]. In SOI technologies this restriction is circumvented by the isolation of the single devices.

Multiple-gate tunneling FETs have been fabricated in a recent low-power MuGFET CMOS technology as presented in Chap. 2 [132]. Since the basic structure of the TFET consists of a p-i-n diode, the implementation of the MuGTFET resembles that of gated diodes used for the bandgap references, see Fig. 4.9. In contrast to the p-i-n diode the gate is not connected to the n^+ implant. To obtain a MuGTFET the layout of a standard device needs to be adapted, whereas the process flow is not modified. The masks for the n^+ and p^+ doping of the active silicon areas are drawn to adjoin in the middle of the gate to obtain complementary source/drain implants. Since the workfunction of the gate is defined by the metal, the n^+ and p^+ implants into the poly on top of the metal layer do not affect the device operation.

5.2 Measurement Results

5.2.1 *I–V Characteristics*

Figure 5.2 shows the measured transfer characteristics of a multi-gate tunneling FET operated in p-type mode (source at n^+-junction, $V_{GS} < 0$, $V_{DS} < 0$). The measured device exhibits a gate length L_{gate} of 65 nm and 8 fins in parallel. The equivalent gate width is 1.2 μm. For $V_{DS} = -0.5$ V the absolute value of the drain current

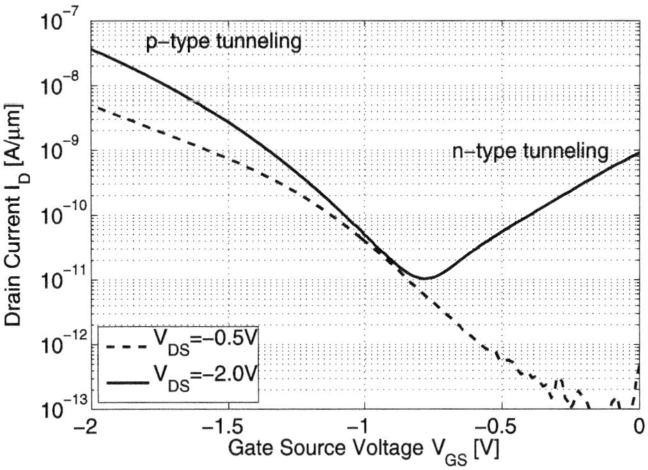

Fig. 5.2 Measured transfer characteristics of MuGTFET operated in p-type mode with $L_{gate} = 65$ nm

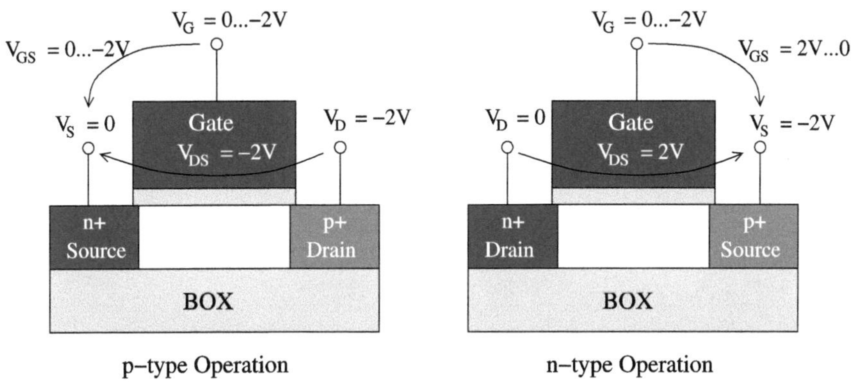

Fig. 5.3 Complementary tunneling within one device: p-type (*left*) and n-type (*right*) operation

rises monotonously with the absolute value of V_{GS}, as expected. Applying a drain-source voltage of -2 V results in different characteristics: starting from very small values of $|V_{GS}|$ the drain current first decreases until it reaches a minimum in the range of $V_{GS} \approx -0.8$ V. With further increasing $|V_{GS}|$, the drain current rises again monotonously. This effect is attributed to the onset of the n-type tunneling current for gate-source voltages smaller than -0.8 V. Lowering the absolute value of $|V_{GS}|$ from -2 V to 0 combined with a drain (for p-type at p^+-junction) potential of -2 V is equivalent to increasing V_{GS} for the n-TFET with source at p^+-junction as explained in Fig. 5.3. Thus complementary tunneling currents are obtained within one device. The slope of drain current is not constant with V_{GS}, which is consistent with theory [133].

5.2.2 Digital and Analog Performance

The measured basic device characteristics are summarized in Table 5.1. The threshold voltage V_T is defined according to a modified constant current criterion as commonly used for MOSFETs. Using this definition, V_T is easily extracted from measurements. Another definition, which is related to the physics behind the TFET is given in [134]: V_T is defined as the gate voltage for which the energy barrier narrowing starts to saturate with the applied gate voltage. Compared to standard MuGFETs in the same technology the achievable on-currents are lowered [6], indicating that the current is limited by the tunneling junction. The off-current is limited by the onset of n-type tunneling, not by the pin-diode leakage. Nevertheless competitive values are obtained. The dependence of on- and off-current for different gate lengths is shown in Fig. 5.4(a). Increasing L_{gate} from 65 nm to 500 nm degrades I_{ON} by about 40%. Applying the same L_{gate} increase to a standard n-type FinFET in the same technology yields in a current degradation of more than 70%. This indicates again

Table 5.1 Measured intrinsic device performance for p-type MuGTFET

Characteristic	Value	Remark/Condition
L_{gate}	65 nm	physical
N_{fins}	8	number of fins
W_{gate}	1.25 μm	$N_{\text{fins}} \cdot (w_{\text{fin}} + 2h_{\text{fin}})$
V_T	−1.41 V	V_{GS} @ $I_D = 0.1$ nA·(W/L), $V_{DS} = -2$ V
I_{ON}	38 nA/μm	$V_{GS} = V_{DS} = -2$ V
I_{OFF}	11 pA/μm	$I_D = \min$, $V_{DS} = -2$ V
Sub-V_T slope	210 mV/Dec	$V_{GS} = V_T - 0.1$ V, $V_{DS} = -2$ V

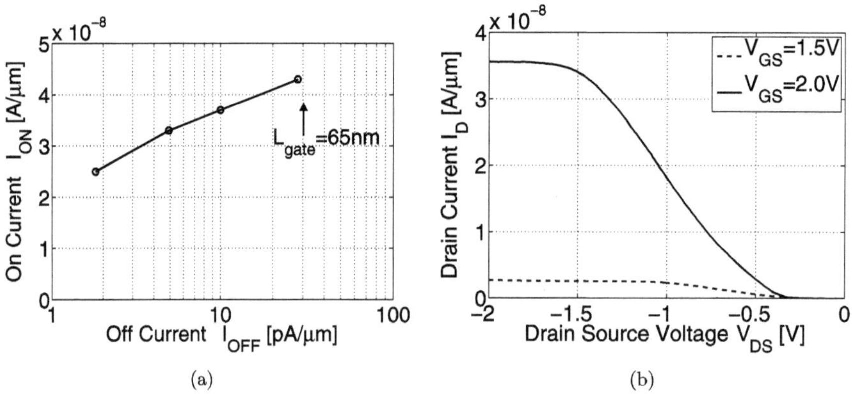

Fig. 5.4 On- versus off-current (**a**) for different channel lengths ($L_{gate} = 65$ nm, 80 nm, 120 nm, 500 nm) and output characteristics of 65 nm p-type MuGTFET (**b**)

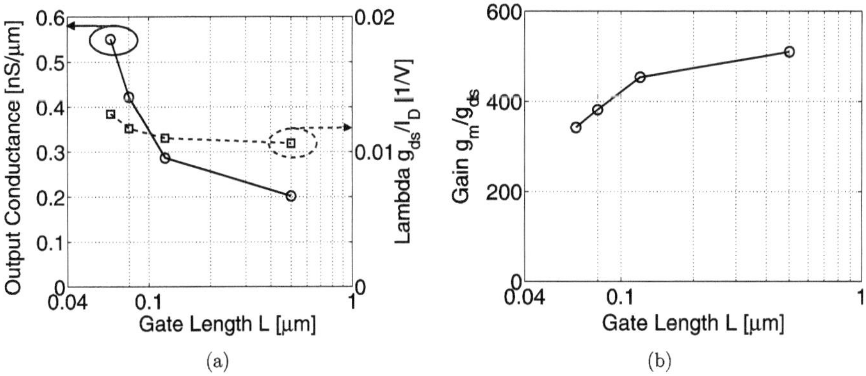

Fig. 5.5 Analog device characteristics of p-MuGTFET for different channel lengths ($V_{GS} = V_{DS} = -2$ V)

that the current of the multi-gate tunneling FET is rather limited by the non-ideal tunneling junction than by the channel resistance.

The output characteristics of a 65 nm pTFET is shown in Fig. 5.4(b). The drain current saturates as expected. Similar to a MOSFET the saturation voltage depends on V_{GS}. A kind of threshold is observed for the drain current for low V_{DS} values, which is also attributed to the non-ideal tunneling junction. For significant current flow additional bending of energy bands due to the applied drain-source voltage is necessary [134].

Analog device performance is investigated by a small signal parameter analysis. g_{ds} and g_m are extracted from measured I–V characteristics. Figure 5.5 shows the resulting values for output conductance and intrinsic gain g_m/g_{ds} for different channel lengths. The multi-gate tunneling FET shows very low values for g_{ds}, i.e. the output resistance is very high. For a fair comparison also the channel length modulation factor $\lambda = g_{ds}/I_D$, i.e. g_{ds} normalized to the drain current is shown. Even for minimum L_{gate} λ is about 0.01 V^{-1}, which is significantly lower than for similar MOSFETs [63]. Moreover λ is only slightly dependent on L_{gate}. Compared to the FinFET devices shown in Chap. 2 the dependence of λ on channel length is reduced by a factor of 3, which proves the scaling potential of the device. Also g_m/g_{ds} is improved compared to MOSFETs: for L_{min} a value of about 340 is obtained, which is more than 3 times higher than reported for comparable FinFET devices. Hence the multi-gate tunneling FET shows promising analog performance. However the yet available currents may be too low for use in standard applications as operational amplifiers that have to drive a load impedance.

5.2.3 Temperature Characteristics

Transfer characteristics of 65 nm multi-gate tunneling FETs operated in p-type mode are measured at temperatures from 10°C up to 120°C, see Fig. 5.6. As ex-

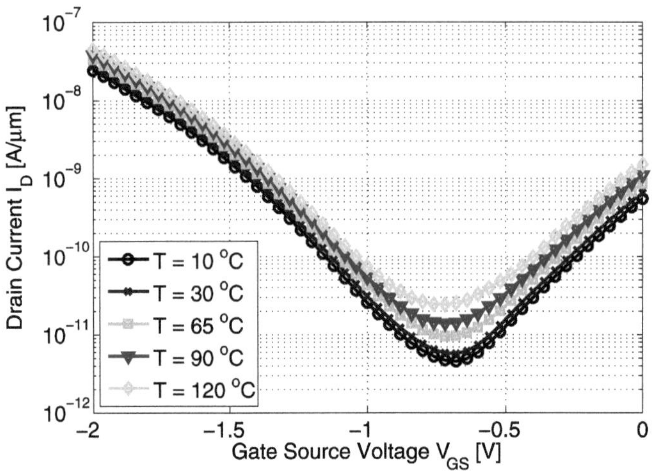

Fig. 5.6 I_D–V_{GS} characteristics of 65 nm p-type MuGTFET at different temperatures

pected from theory [135] and prior work [136], the tunneling current shows a weak, positive temperature dependence. This can be explained with a simple approximation of the band-to-band tunneling current as function of the electrical field F [135]

$$I_{B2B} \propto A \cdot F^{5/2} \cdot \exp\left(-\frac{B}{F}\right). \tag{5.1}$$

The fitting parameters are given by $A \propto E_G^{-7/4}$ and $B \propto E_G^{3/2}$ with the bandgap energy E_G. Thus, the temperature dependence of the tunneling current results mainly from the weak temperature dependence of E_G. Extracted temperature coefficients of the absolute values of drain current and gate-source voltage are 1.87×10^{-10} A/°K and -0.92 mV/°K, respectively. The low but well defined dependence of the I–V characteristics on temperature offers potential benefits for reference circuits.

5.2.4 Variations

For a first assessment of variation and mismatch effects, the transfer characteristics of different 65 nm multi-gate tunneling FETs operated in p-type mode are measured over a whole 200 mm wafer. V_T is extracted for all devices. The statistical distribution of V_T is shown in Fig. 5.7(a). With a mean value of -1.497 V the relative standard deviation of V_T is 2.6% which is comparable to standard MOSFET devices in similar technology [71]. From these first investigations no additional variations are identified originating from the tunneling junction or the modified device layout. The adjoining masks for n^+ and p^+ implants over the gate introduce no additional variations of V_T because the workfunction of the gate electrode material determines

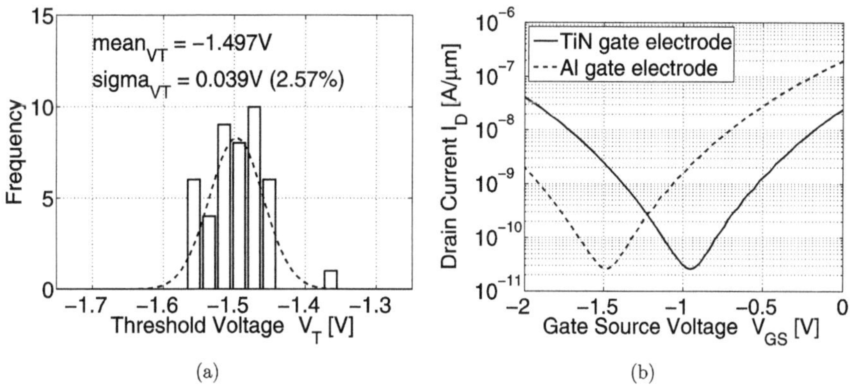

Fig. 5.7 V_T distribution over whole wafer ($V_{DS} = -2.0$ V) and simulated I_D–V_{GS} characteristics of 65 nm p-MuGTFET with TiN and Aluminum gate electrode at $V_{DS} = -2.0$ V

V_T in the fully depleted devices. This is proven for MuGTFETs by device simulation.

5.3 Device Simulation

Device simulations are performed to gain a deeper understanding of the parameters that determine the device behavior and to derive optimization strategies for the multi-gate tunneling FET. The numerical device simulator Sentaurus is used. A more detailed description of the simulation approach is given in [137] and [138]. Tunneling currents between strongly tilted energy bands are implemented in the drift-diffusion-like transport model by adding equivalent generation rates to the carrier source terms in the balance equations, which represent the generation of electron-hole pairs in the middle of the bandgap. Phonon-assisted tunneling is modeled on the basis of the Kubo formalism for the tunneling conductivity. Due to the large computational effort the three-dimensional device geometry is approximated by a cylindrical realization.

Simulated I_D–V_{GS} characteristics for devices with TiN and aluminum gate electrode material are shown in Fig. 5.7(b). Two important conclusions can be drawn from these results: By numerical device simulation it can be proven that the current increase for low V_{GS} is caused indeed by n-type band-to-band tunneling at the p^+-side and is no measurement artefact. In addition the whole I_D–V_{GS} curve can be shifted by changing the gate electrode material (i.e. its workfunction). Thus, the V_T of the MuGTFET can be adjusted in the same way as the V_T of MOSFETs, whereas the physics defining the corresponding threshold is different. Shifting the leakage minimum in the I_D–V_{GS} curve of n- and pTFET close to zero, i.e. suppressing the respective parasitic tunneling currents using proper gate materials is one approach for device optimization.

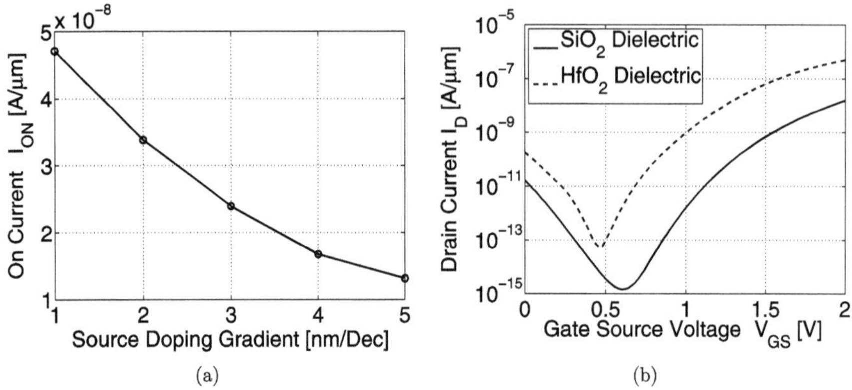

Fig. 5.8 Simulated on-current for varying doping gradients (**a**) and I_D–V_{GS} curves for different gate dielectric materials (n-type operation)

Next the impact of the doping profiles on the tunneling currents is investigated. In Fig. 5.8(a) simulated on-currents for different doping gradients of the source implant are depicted. The current is significantly increased for steeper doping gradients, similar to planar and vertical TFETs [35, 139]. Steeper doping profiles result in stronger band bending and enhanced tunneling probability, resulting in higher on-currents. Moreover a steeper slope of the I_D–V_{GS} characteristics and lower off-currents can be observed. The dependence of tunneling current on doping gradients enables also an optimization of leakage currents: using a smoother doping profile at the drain side of the device weakens the complementary tunneling junction and results in suppressed parasitic leakage currents. Consequently, optimizing the doping profiles at the source/drain junctions is a strong lever for improving multi-gate tunneling FET performance. The strong dependence of the tunneling current on the source doping gradient explains the low currents of MuGTFET in a standard FinFET process compared to planar TFETs [140]. The tilted and twisted implant process of the LDD extension in the three-dimensional fin structure results in lower doping gradients compared to planar devices.

Since the band-to-band tunneling current shows an exponential dependence on the electric field from gate to source, another optimization possibility is the use of thinner gate oxides. Keeping the gate and drain bias constant the field at the tunneling junction is increased, resulting in higher tunneling rates and on-currents. Alternatively high-k dielectrics could be used, which will lead to the same effect, whereas the oxide thickness remains constant and gate leakage currents due to direct tunneling are suppressed. Figure 5.8(b) illustrates the consequences on the transfer characteristics changing from conventional SiO_2 dielectric to HfO_2. Using the high-k dielectric increases the drain current up to two orders of magnitude, depending on V_{GS}.

5.4 MuGTFET Reference Circuit

Due to the low currents the use of multi-gate TFETs in standard applications as digital logic gates or amplifiers seems difficult in current standard technologies. Process optimization is necessary to achieve competitive performance. However the MuGTFET offers some interesting features for special applications already today. Especially the negative differential resistance (NDR), i.e. the part of the MuGTFET $I-V$ characteristics with decreasing current for increasing $|V_{GS}|$ could be used, e.g. in low-power voltage controlled oscillators. In this work we present a voltage reference circuit as possible application for the MuGTFET. Voltage references are intended to deliver a well defined output voltage, independent of supply voltage, temperature and process variations.

The schematic of the proposed reference circuit and the idea behind is shown in Fig. 5.9. Standard MuGFETs and MuGTFETs are combined in this circuit. The FinFETs $M_{1...3}$, the MuGTFET T_1 and the resistor R comprise a current reference circuit, similar to the CMOS reference in Fig. 4.3. The reference current through M_2 is in first order only determined by the resistor value and gate-source voltage of MuGTFET T_1, but independent of the supply voltage. Since V_{GS} of T_1 equals the voltage drop over the resistor, only intersections of the linear resistor characteristics with the exponential TFET characteristics are valid operating points of the circuit, see Fig. 5.9. Consequently the reference current decreases proportional with temperature, realizing an Inverse Proportional to Absolute Temperature (IPTAT) current source. The IPTAT current source part of the reference circuit could be used independently as temperature sensor. The decreasing current is forced into the MOSFET M_5 via the current mirror M_2–M_3. M_5 is operated in a bias point where mobility degradation is dominating the temperature dependence of the drain current. So the drain current of M_5 also decreases with temperature. With proper choice of the corresponding W/L ratios, the gate-source voltage of M_5 stays constant over temperature and can be used as reference voltage as shown in Fig. 5.9. In theory the

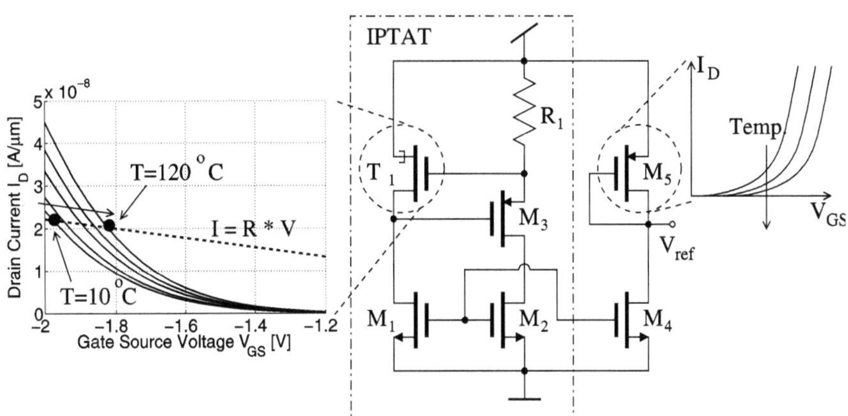

Fig. 5.9 Schematic of the proposed voltage reference circuit employing multi gate TFETs and standard FinFETs

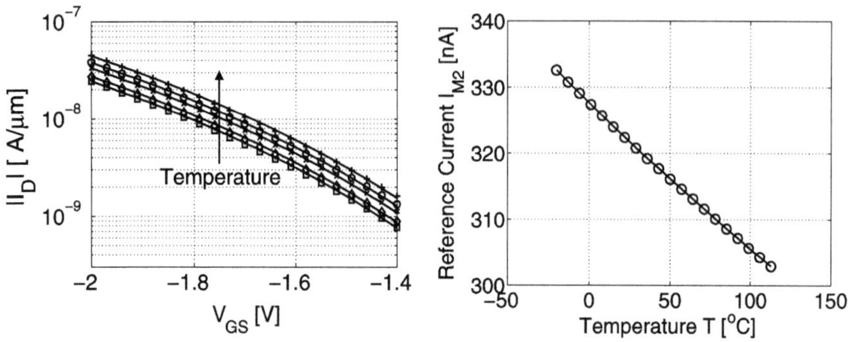

(a) (b)

Fig. 5.10 Simulation (*lines*) and measurement (*symbols*) of I_D–V_{GS} characteristics for different temperatures and simulated IPTAT current (*right*)

IPTAT current source would also work with a MOSFET operated in sub-threshold region, where the drain current has a positive temperature coefficient. However this temperature dependence significantly changes with increasing V_{GS}. In contrast the temperature dependence of the tunneling current is almost constant for a large range of bias points. So the TFET circuit is less sensitive to variations and mismatch. To verify the concept by circuit simulation a VHDL-AMS model of the TFET was developed. A modified version of Hurkx's equation with fitting parameters C_1, C_2, bandgap energy E_G and temperature T is used:

$$I_D = C_1 \cdot E_G^{-1/2} \cdot T \cdot V_{GS}^2 \cdot \exp\left(\frac{-C_2 \cdot E_G^{3/2}}{V_{GS}}\right). \tag{5.2}$$

A comparison of measured and simulated device characteristics in Fig. 5.10(a) shows good agreement.

To meet todays requirements on a low supply voltage, the V_T of the measured devices is reduced by 0.8 V in the model, assuming that an appropriate gate material is available.

Using the VHDL-AMS model for the multi-gate tunneling FETs and PSP compact models for the standard FinFETs the complete reference circuit is optimized and simulated in SPICE. The IPTAT current is shown in Fig. 5.10(b). The resulting dependence of the reference voltage V_{ref} on temperature is shown in Fig. 5.11(a). A very low temperature coefficient of 37 ppm V/°K is obtained for V_{ref}. Variations of the supply voltage V_{DD} also have low impact on V_{ref}. The simulation shown in Fig. 5.11(b) reveals a power supply rejection ratio of more than 30 dB. The power consumption is about 10 μW at $V_{DD} = 1.5$ V.

Summing up, the integration of tunneling FETs in a low-power MuGFET technology is feasible, nevertheless the achievable on-currents are significantly lower than in comparable MOSFETs. However promising analog properties and low variability regarding temperature and threshold voltage are demonstrated. Device simulations prove the scalability and suggest a tuning of doping profiles and gate stack

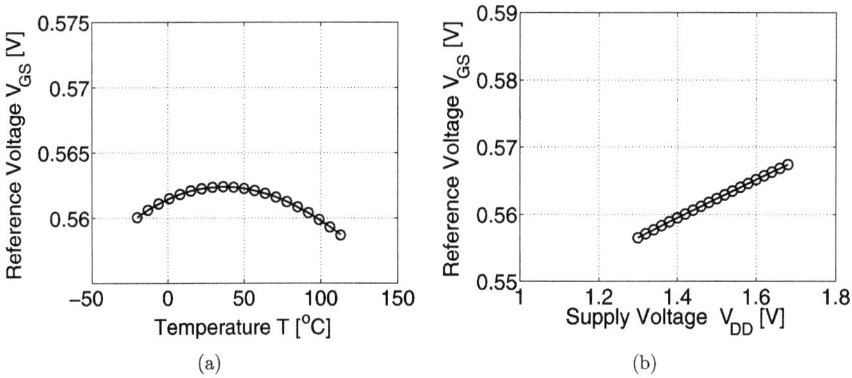

Fig. 5.11 Simulated temperature and V_{DD} dependence of V_{ref}

for optimization. A novel reference circuit using TFETs and MOSFETs is proposed, robust against temperature and supply voltage variations.

Chapter 6
Conclusions and Outlook

In this work analog and mixed-signal circuit design aspects in emerging multi-gate CMOS technologies have been discussed, focused on performance and variability aspects. FinFETs employing high-k gate dielectrics are considered as promising scaling scenario overcoming the scaling limitations of conventional planar bulk CMOS. The feasibility of analog and mixed-signal circuits in these technologies is investigated on device and circuit level, where multi-gate and high-k related aspects are separated. Thus, the results corresponding to high-k related effects are applicable also to advanced planar bulk CMOS employing high-k dielectrics. As far as possible FinFET is benchmarked against planar bulk.

Besides process complexity, the main concerns with respect to the introduction of high-k are related to defects in the dielectric layers and to degraded quality of interfaces yielding a high amount of traps. Both aspects can cause higher flicker noise. Another phenomenon corresponding to pronounced charge trapping is the dynamic variation of threshold voltage or threshold hysteresis. To analyze the impact of this new effect on circuit level a scalable simulation model is developed and fitted to measurements. Degraded reliability, e.g. bias temperature instabilities is another aspect relevant for analog and RF, which needs to be considered in future work.

From circuit point of view increased flicker noise and dynamic V_T variations present serious challenges. The impact of flicker noise on the overall noise budget is determined by the system bandwidth. Reduction of noise is always connected with power and area penalty. Dedicated noise reduction techniques are useful in some cases but introduce other issues like switching noise. Improved switched bias techniques using forward body bias seem promising, nevertheless they are not applicable in fully depleted multi-gate technologies.

A systematic analysis of transient V_T variations reveals that hysteresis effects in general are no show stopper for analog and mixed-signal circuit design as long as the maximum V_T shift does not exceed $2\ldots3$ mV. However, comparators are very sensitive against hysteresis effects, even for small V_T shifts incorrect decisions can occur. In worst case the comparator resolution is limited to the maximum V_T shift. If applicable, conventional offset compensation reduces the effect, comparators with switched input pairs are proposed as alternative. Flash ADCs typically are not affected by charge trapping induced comparator errors due to their limited resolution.

M. Fulde, *Variation Aware Analog and Mixed-Signal Circuit Design in Emerging Multi-Gate CMOS Technologies*, Springer Series in Advanced Microelectronics 28,
DOI 10.1007/978-90-481-3280-5_6, © Springer Science+Business Media B.V. 2010

Also in single-bit $\Sigma\Delta$ ADCs comparator errors are no serious concern, since they are suppressed in the same way as quantization noise. However, SAR A/D converters strongly suffer even from small hysteresis effects, since the overall resolution is limited to the comparator resolution and standard offset compensation is not applicable in this case. Simulations and measurement results show that besides comparators with switched input pairs the use of redundancy by means of non-binary search represents an efficient countermeasure. Future work should address the impact of hysteresis on very high resolution converters like high order multi-bit $\Sigma\Delta$ ADCs. Other aspects to be analyzed are wear out effects, i.e. increase of hysteresis level during lifetime and statistical mismatch of hysteresis parameters.

On device level the main advantage of fully depleted FinFETs compared to planar bulk devices is the possibility to control short channel effects without high channel doping and HALO implants. Reduced short channel effects and missing HALO implants significantly improve output impedance and intrinsic voltage gain. The undoped channel region also reduces random doping fluctuation and yields better V_T matching. Since sufficient short channel control of thin body, fin or nanowire FETs always requires a thin (small) silicon film (volume) compared to the gate length, these devices inherently have a high series resistance with reduced g_m and drive current. In case of the FinFET SEG was shown to efficiently reduce series resistance, however at the cost of increased parasitic capacitance. Although f_t and f_{max} of state-of-the-art FinFETs reach the 150 GHz regime and enable RF applications up to 10 GHz, comparable planar bulk devices still show better performance.

Important analog, mixed-signal and RF building blocks are implemented in a multi-gate circuit assessment. Current mirror based matching test-structures show that new mismatch effects related to the fin structure have to be considered in layout. The beneficial analog properties, i.e. improved output resistance and intrinsic gain directly translate into improved circuit performance: FinFET current references are much less sensitive against variations of output or supply voltage. The high intrinsic gain significantly improves the open loop gain of OpAmps. The FinFET implementation of a two stage Miller OpAmp achieves more than 65 dB open loop gain, about 15 dB more than the planar counterpart. In some cases the improved open-loop gain can be traded against lower power consumption or higher gain-bandwidth. The feasibility of low voltage bandgap references using gated p-i-n diodes is proven by measurements. The p-i-n diode, which is available in MuGFET processes without extra costs, contributes no additional variations to the reference voltage. Low g_{ds} of current source transistors, low Op-Amp offset and sufficient gain are identified as main parameters for optimization of multi-gate bandgap performance. A 10 bit current steering DAC is used to demonstrate that beneficial analog and matching behavior of FinFETs enable a reduction of circuit area without performance degradation. Competitive jitter performance per area and power consumption is achieved with a simple analog charge pump FinFET PLL. No degradation of flicker or thermal device noise is observed. The characterization of a 2 GHz LNA confirms this statement. The degradation of phase noise in a 2 GHz LC-VCO is mainly attributed to the non-optimized BEOL and varactors, presenting no general device related limit.

Self heating is discussed as SOI specific effect leading to transient parameter variations. Due to the low power density in typical analog applications self heating has to be considered only in exceptional cases. Thermal coupling can be optimized by dedicated layout measures. Simulations indicate an increase of self-heating with scaling of fin dimensions, i.e. active silicon volume. Thus self heating may gain importance in further scaled technology nodes. Although short channel control dictates the use of minimum fin width in digital circuits, the fin width could be seen as additional design parameter for analog and RF. Stacked current sources and multi fin width devices with improved VIP3 are shown as example to demonstrate how selective adaption of fin width yields improved circuit performance.

From analog point of view the superior impedance and gain properties as well as the beneficial matching behavior are strong arguments for the introduction of FinFETs in post 32 nm nodes. FinFETs overcome two of the major analog concerns in scaled CMOS: low intrinsic gain and high variability. The results shown here prove that the use of FinFET enables improved circuit performance or decreased area. RF applications in the mm-wave range are still very challenging due to the inherently increased parasitics. Further FinFET integration challenges regarding 22 nm and below are fin patterning, implementation of strain and contacting of single fins without landing pad.

Finally multi-gate tunneling FETs are discussed as outlook for analog design issues in post CMOS technologies. The integration of tunneling FETs in a low-power MuGFET technology is demonstrated. Digital and analog device characteristics show low dependence on channel length, scaling of the MuGTFET to channel lengths in the 10 nm regime seems feasible. Due to non-optimized doping profiles and gate stack the on-currents are still lower than in comparable MOSFETs. Device simulations show that gate stack engineering and source/drain doping profile tuning can be used to optimize device characteristics for circuit applications. The realized MuGTFETs show promising analog device performance. No additional variability issues are identified from analysis of temperature dependence and matching behavior. A new voltage reference circuit is proposed and verified with simulations. The circuit features low dependence on supply voltage and temperature.

Symbols and Abbreviations

A_0	DC open loop voltage gain
A_{V_T}	Matching constant for threshold voltage
A_β	Matching constant for current factor
β	Current factor
BW	Bandwidth
C_C	Compensation capacitance
C_{GD}	Gate-drain capacitance
C_{GS}	Gate-source capacitance
C_L	Load capacitance
C_{ox}	Specific oxide capacitance
C_{par}	Parasitic capacitance
C_{th}	Thermal capacitance in self heating model
ϵ_r	Dielectric constant
E_G	Bandgap energy
ef	Exponent of flicker noise frequency dependence
F	Electric field
f	Frequency
f_{ain}	Analog input frequency
f_{chop}	Chopping frequency
f_{max}	Maximum oscillation frequency
f_{nd}	Non-dominant pole frequency
f_{sample}	Sampling frequency
f_t	Transit frequency
g_{ds}	Small signal output conductance
g_m	Small signal transconductance
h_{fin}	Fin height
I_D	Drain current
I_{B2B}	Band-to-band tunneling current
I_{GATE}	Gate current at $V_{GS} = 0$, $V_{DS} = V_{DD}$
I_{ON}	On current at $V_{GS} = V_{DS} = V_{DD}$
I_{OFF}	Off current at $V_{GS} = 0$, $V_{DS} = V_{DD}$

M. Fulde, *Variation Aware Analog and Mixed-Signal Circuit Design in Emerging Multi-Gate CMOS Technologies,* Springer Series in Advanced Microelectronics 28, DOI 10.1007/978-90-481-3280-5, © Springer Science+Business Media B.V. 2010

I_S	Diode saturation current
κ	Dielectric constant
K_f	Flicker noise coefficient
k_i	Scaling coefficient i of time constants in V_{shift} model
L_{el}	Effective channel length
L or L_{gate}	Transistor gate length
L_{min}	Minimum transistor gate length
μ	Charge carrier mobility
n	Number of bits
N	Noise power
N_A	Doping level
P	Power
$\Phi_{1/2}$	Complementary, non-overlapping clock phases
R_{DD}	Sensitivity of output current against supply voltage variations
R_G	Gate resistance
R_{Leak}	Diode leakage resistance
R_{out}	Output resistance
R_{SD}	Source-drain series resistance
R_{scale}	Scaling resistance in V_{shift} model
R_{th}	Thermal resistance in self heating model
s	Scaling factor
S	Signal power
T	Temperature
t_{dep}	Penetration depth of gate field
t_{eval}	Evaluation time
t_{ox}	Oxide thickness
t_{si}	Thickness of silicon layer
t_{stress}	Stress time
V_{bi}	Build in source potential
V_{CM}	Common mode voltage
V_{DD}	Positive supply voltage
V_D	Diode voltage
V_{DS}	Drain-source voltage
V_{GS}	Gate-source voltage
V_{ov}	Overdrive voltage $V_{GS}-V_T$
$V_{ref(p)(n)}$	(Positive) (negative) reference voltage
V_T	Threshold voltage
V_{the}	Thermal voltage
V_{shift}	Shift voltage representing time dependent V_T shift
V_{SS}	Negative supply voltage
V_{ss}	Maximum steady state V_T shift
V_{stress}	Stress voltage
w_{fin}	Fin width
W	Transistor (gate) width
x_j	Junction depth

ADC	Analog-to-digital converter
BEOL	Back end of line (metallization)
BGR	Bandgap reference
BOX	Buried oxide layer
CMRR	Common mode rejection ratio
DAC	Digital-to-analog converter
DIBL	Drain induced barrier lowering
DNL	Differential non-linearity
EI	Electrostatic integrity
ENOB	Effective number of bits
EOT	Equivalent oxide thickness
FD	Fully depleted
FET	Field effect transistor
GBW	Gain bandwidth product
INL	Integral non-linearity
IPTAT	Inverse proportional to absolute temperature
LDD	Lightly doped drain
LNA	Low noise amplifier
LSB	Least significant bit
MuGFET	Multi-gate FET
MuGTFET	Multi-gate tunneling FET
NDR	Negative differential resistance
NTF	Noise transfer function
OPC	Optical proximity correction
OSR	Oversampling ratio
PLL	Phase locked loop
PSRR	Power supply rejection ratio
SAR	Successive approximation register
SCE	Short channel effect
SEG	Selective epitaxial growth
SFDR	Spurious free dynamic range
SIP	System in package
SNDR	Signal-to-noise-and-distortion ratio
SNR	Signal-to-noise ratio
SOC	System on chip
SOI	Silicon on insulator
SR	Slew rate
(P)(N)TC	(Positive) (negative) temperature coefficient
TFET	Tunneling FET
THD	Total harmonic distortion
VCO	Voltage controlled oscillator
VIP3	Third order interception point
VTRO	Threshold voltage roll-off

References

1. T. Luftner, J. Berthold, C. Pacha, G. Georgakos, G. Sauzon, O. Homke, J. Beshenar, P. Mahrla, K. Just, P. Hober, S. Henzler, D. Schmitt-Landsiedel, A. Yakovleff, A. Klein, R. Knight, P. Acharya, H. Mabrouki, G. Juhoor, M. Sauer, A 90 nm CMOS low-power GSM/EDGE multimedia-enhanced baseband processor with 380 MHz ARM9 and mixed-signal extensions, in *IEEE International Solid-State Circuits Conference, Digest of Technical Papers*, 6–9 February 2006, pp. 952–961
2. M. Hammes, C. Kranz, J. Kissing, D. Seippel, P.-H. Bonnaud, E. Pelos, A GSM baseband radio in 0.13 um CMOS with fully integrated power-management, in *IEEE International Solid-State Circuits Conference, Digest of Technical Papers* (2007), pp. 264–602
3. S. Heinen, Architectures and circuit techniques for nanoscale RF CMOS, in *ISSCC Advanced Design Forum* (2008)
4. International Technology Roadmap for Semiconductors, http://www.itrs.net, 2007 edn. Chap. System Drivers
5. B.S. Meyerson, Opening keynote address—how does one define "technology" now that classical scaling is dead (and has been for years)? in *Proceedings of 42nd Design Automation Conference*, June 2005, p. 25
6. K. von Arnim, E. Augendre, C. Pacha, J. Berthold, T. Schulz, K.T. San, F. Bauer, A. Nackaerts, R. Rooyackers, N. Collaert, T. Vandeweyer, B. Degroote, A. Dixit, R. Siganamalla, W. Xiong, A. Marshall, C.-R. Cleavelin, K. Schruefer, M. Jurczak, A low-power multi-gate FET CMOS technology with 13.9 ps inverter delay, large-scale integrated high performance digital circuits and SRAM, in *Symposium on VLSI Technology, Digest of Technical Papers* (2007), pp. 106–107
7. C. Pacha, K. von Arnim, F. Bauer, T. Schulz, W. Xiong, K.T. San, A. Marshall, T. Baumann, C.-R. Cleavelin, K. Schruefer, J. Berthold, Efficiency of low-power design techniques in multigate FET CMOS circuits, in *Proceedings of 33th European Solid-State Circuits Conference, ESSCIRC* (2007), pp. 111–114
8. G. Knoblinger, F. Kuttner, A. Marshall, C. Russ, P. Haibach, P. Patruno, T. Schulz, W. Xiong, M. Gostkowski, K. Schruefer, C.R. Cleavelin, Design and evaluation of basic analog circuits in an emerging MuGFET technology, in *2005 IEEE International SOI Conference Proceedings* (IEEE Press, New York, 2005), pp. 39–40
9. International Technology Roadmap for Semiconductors, http://www.itrs.net, 2007 edn. Chap. Process Integration, Devices, and Structures
10. R. Khamankar, H. Bu, C. Bowen, S. Chakravarthi, P.R. Chidambaram, M. Bevan, A. Krishnan, H. Niimi, B. Smith, J. Blatchford, B. Hornung, J.P. Lu, P. Nicollian, B. Kirkpatrick, D. Miles, M. Hewson, D. Farber, L. Hall, H. Alshareef, A. Varghese, A. Gurba, V. Ukraintsev, B. Rathsack, J. DeLoach, J. Tran, C. Kaneshige, M. Somervell, S. Aur, C. Machala, T. Grider, An enhanced 90 nm high performance technology with strong performance improvements from

M. Fulde, *Variation Aware Analog and Mixed-Signal Circuit Design in Emerging Multi-Gate CMOS Technologies,* Springer Series in Advanced Microelectronics 28, DOI 10.1007/978-90-481-3280-5, © Springer Science+Business Media B.V. 2010

stress and mobility increase through simple process changes, in *Symposium on VLSI Technology, Digest of Technical Papers*, June 2004, pp. 162–163

11. A. Steegen, R. Mo, R. Mann, M.-C. Sun, M. Eller, G. Leake, D. Vietzke, A. Tilke, F. Guarin, A. Fischer, T. Pompl, G. Massey, A. Vayshenker, W.L. Tan, A. Ebert, W. Lin, W. Gao, J. Lian, J.-P. Kim, P. Wrschka, J.-H. Yang, A. Ajmera, R. Knoefler, Y.-W. Teh, F. Jamin, J.E. Park, K. Hooper, C. Griffin, P. Nguyen, V. Klee, V. Ku, C. Baiocco, G. Johnson, L. Tai, J. Benedict, S. Scheer, H. Zhuang, V. Ramanchandran, G. Matusiewicz, Y.-H. Lin, Y.K. Siew, F. Zhang, L.S. Leong, S.L. Liew, K.C. Park, K.-W. Lee, D.H. Hong, S.-M. Choi, E. Kaltalioglu, S.O. Kim, M. Naujok, M. Sherony, A. Cowley, A. Thomas, J. Sudijohno, T. Schiml, J.-H. Ku, I. Yang, 65 nm CMOS technology for low power applications, in *Technical Digest of International Electron Devices Meeting, IEDM*, December 2005, pp. 64–67
12. F. Arnaud, B. Duriez, B. Tavel, L. Pain, J. Todeschini, M. Jurdit, Y. Laplanche, F. Boeuf, F. Salvetti, D. Lenoble, J.P. Reynard, F. Wacquant, P. Morin, N. Emonet, D. Barge, M. Bidaud, D. Ceccarelli, P. Vannier, Y. Loquet, H. Leninger, F. Judong, C. Perrot, I. Guilmeau, R. Palla, A. Beverina, V. DeJonghe, M. Broekaart, V. Vachellerie, R.A. Bianchi, B. Borot, T. Devoivre, N. Bicais, D. Roy, M. Denais, K. Rochereau, R. Difrenza, N. Planes, H. Brut, L. Vishnobulta, D. Reber, P. Stolk, M. Woo, Low cost 65 nm CMOS platform for low power & general purpose applications, in *Symposium on VLSI Technology, Digest of Technical Papers*, June 2004, pp. 10–11
13. M. Iwai, A. Oishi, T. Sanuki, Y. Takegawa, T. Komoda, Y. Morimasa, K. Ishimaru, M. Takayanagi, K. Eguchi, D. Matsushita, K. Muraoka, K. Sunouchi, T. Noguchi, 45 nm CMOS platform technology (CMOS6) with high density embedded memories, in *Symposium on VLSI Technology, Digest of Technical Papers*, June 2004, pp. 12–13
14. F. Boeuf, F. Arnaud, M.T. Basso, D. Sotta, F. Wacquant, J. Rosa, N. Bicais-Lepinay, H. Bernard, J. Bustos, S. Manakli, M. Gaillardin, J. Grant, T. Skotnicki, B. Tavel, B. Duriez, M. Bidaud, P. Gouraud, C. Chaton, P. Morin, J. Todeschini, M. Jurdit, L. Pain, V. De-Jonghe, R. El-Farhane, S. Jullian, A conventional 45 nm CMOS node low-cost platform for general purpose and low power applications, in *Technical Digest of IEEE International Electron Devices Meeting, IEDM*, December 2004, pp. 425–428
15. V. Misra, G. Lucovsky, G. Parsons, Issues in high-k gate stack interfaces. MRS Bull. **27**, 212–216 (2001)
16. H.R. Huff, A. Hou, C. Lim, Y. Kim, J. Barnett, G. Bersuker, G.A. Brown, C.D. Young, P.M. Zeitzoff, J. Gutt, P. Lysaght, M.I. Gardner, R.W. Murto, High-k gate stacks for planar, scaled CMOS integrated circuits. Microelectron. Eng. **69**, 152–167 (2003)
17. C. Hobbs, L. Fonseca, V. Dhandapani, S. Samavedam, B. Taylor, J. Grant, L. Dip, D. Triyoso, R. Hegde, D. Gilmer, R. Garcia, D. Roan, L. Lovejoy, R. Rai, L. Hebert, H. Tseng, B. White, P. Tobin, Fermi level pinning at the poly-Si/metal oxide interface, in *Symposium on VLSI Technology, Digest of Technical Papers*, June 2003, pp. 9–10
18. R. Chau, S. Datta, M. Doczy, B. Doyle, J. Kavalieros, M. Metz, High-k/metal-gate stack and its MOSFET characteristics. IEEE Electron Device Lett. **25**(6), 408–410 (2004)
19. Q. Lu, Y. Yeo, P. Ranade, H. Takeuchi, T. King, Ch. Hu, S.C. Song, H.F. Luan, D. Kwong, Dual-metal gate technology for deep-submicron CMOS transistors, in *Symposium on VLSI Technology, Digest of Technical Papers* (2000), pp. 72–73
20. K. Mistry, C. Allen, C. Auth, B. Beattie, D. Bergstrom, M. Bost, M. Brazier, M. Buehler, A. Cappellani, R. Chau, C.-H. Choi, G. Ding, K. Fischer, T. Ghani, R. Grover, W. Han, D. Hanken, M. Hattendorf, J. He, J. Hicks, R. Huessner, D. Ingerly, P. Jain, R. James, L. Jong, S. Joshi, C. Kenyon, K. Kuhn, K. Lee, H. Liu, J. Maiz, B. Mclntyre, P. Moon, J. Neirynck, S. Pae, C. Parker, D. Parsons, C. Prasad, L. Pipes, M. Prince, P. Ranade, T. Reynolds, J. Sandford, L. Shifren, J. Sebastian, J. Seiple, D. Simon, S. Sivakumar, P. Smith, C. Thomas, T. Troeger, P. Vandervoorn, S. Williams, K. Zawadzki, A 45 nm logic technology with high-k+metal gate transistors, strained silicon, 9 Cu interconnect layers, 193 nm dry patterning, and 100% Pb-free packaging, in *Technical Digest of International Electron Devices Meeting, IEDM*, December 2007, pp. 247–250

21. T. Skotnicki, G. Merckel, T. Pedron, The voltage-doping transformation: a new approach to the modeling of MOSFET short-channel effects. IEEE Electron Device Lett. **9**(3), 109–112 (1988)
22. T. Skotnicki, Heading for decananometer CMOS—is navigation among icebergs still a viable strategy? in *Proceedings of the 30th European Solid-State Device Research Conference, ESSDERC*, September 2000, pp. 19–33
23. J.L. Pelloie, A.J. Auberton-Herv, C. Raynaud, O. Faynot, SOI technology performance and modelling, in *IEEE International Solid-State Circuits Conference, Digest of Technical Papers* (1999), pp. 428–429
24. T. Sekigawa, Y. Hayashi, Calculated threshold-voltage characteristics of an XMOS transistor having an additional bottom gate. Solid-State Electron. **27**, 827–828 (1984)
25. J.P. Denton, G.W. Neudeck, Fully depleted dual-gated thin-film SOI P-MOSFETs fabricated in SOI islands with an isolated buried polysilicon backgate. IEEE Electron Device Lett. **17**(11), 509–511 (1996)
26. D. Hisamoto, T. Kaga, Y. Kawamoto, E. Takeda, A fully depleted lean-channel transistor (DELTA)-a novel vertical ultra thin SOI MOSFET, in *Technical Digest of International Electron Devices Meeting, IEDM*, December 1989, pp. 833–836
27. D. Hisamoto, W.-C. Lee, J. Kedzierski, H. Takeuchi, K. Asano, C. Kuo, E. Anderson, T.-J. King, J. Bokor, C. Hu, FinFET-a self-aligned double-gate MOSFET scalable to 20 nm. IEEE Trans. Electron Devices **47**(12), 2320–2325 (2000)
28. B. Doyle, B. Boyanov, S. Datta, M. Doczy, S. Hareland, B. Jin, J. Kavalieros, T. Linton, R. Rios, R. Chau, Tri-gate fully-depleted CMOS transistors: fabrication, design and layout, in *Symposium on VLSI Technology, Digest of Technical Papers*, June 2003, pp. 133–134
29. J.-T. Park, J.-P. Colinge, C.H. Diaz, Pi-gate SOI MOSFET. IEEE Electron Device Lett. **22**(8), 405–406 (2001)
30. F.-L. Yang, H.-Y. Chen, F.-C. Chen, C.-C. Huang, C.-Y. Chang, H.-K. Chiu, C.-C. Lee, C.-C. Chen, H.-T. Huang, C.-J. Chen, H.-J. Tao, Y.-C. Yeo, M.-S. Liang, C. Hu, 25 nm CMOS Omega FETs, in *Technical Digest of International Electron Devices Meeting, IEDM* (2002), pp. 255–258
31. J.P. Colinge, *FinFETs and Other Multi-Gate Transistors* (Springer, Berlin, 2008)
32. S.-H. Oh, D. Monroe, J.M. Hergenrother, Analytic description of short-channel effects in fully-depleted double-gate and cylindrical, surrounding-gate MOSFETs. IEEE Electron Device Lett. **21**(9), 445–447 (2000)
33. International Technology Roadmap for Semiconductors, http://www.itrs.net, 2007 edn. Chap. Emerging Research Devices
34. W. Fischer, Field induced tunnel diode. IBM Tech. Dis. Bull. **16**(7), 2303 (1973)
35. P.-F. Wang, Th. Nirschl, D. Schmitt-Landsiedel, W. Hansch, Simulation of the Esaki-tunneling FET. Solid State Electron. **47**, 1187–1192 (2003)
36. W.M. Reddick, G.A.J. Amaratunga, Silicon surface tunnel transistor. Appl. Phys. Lett. **67**(4), 494–496 (1995)
37. W. Hansch, C. Fink, J. Schulze, I. Eisele, A vertical MOS-gated Esaki tunneling transistor in silicon. Thin Solid Films **369**, 387–389 (2000)
38. P. Kinget, M. Steyaert, Impact of transistor mismatch on the speed-accuracy-power trade-off of analog CMOS circuits, in *Proceedings of the IEEE Custom Integrated Circuits Conference, CICC 1996* (1996), pp. 333–336
39. P. Gray, P. Hurst, S. Lewis, R. Meyer, *Analysis and Design of Analog Integrated Circuits*, 4th edn. (Wiley, New York, 2001)
40. P. Wambacq, W. Sansen, *Distortion Analysis of Analog Integrated Circuits* (Kluwer Academic, Dordrecht, 1998)
41. J.A. Croon, M. Rosmeulen, S. Decoutere, W. Sansen, H.E. Maes, An easy-to-use mismatch model for the MOS transistor. IEEE J. Solid-State Circuits **37**(8), 1056–1064 (2002)
42. M.J.M. Pelgrom, A.C.J. Duinmaijer, A.P.G. Welbers, Matching properties of MOS transistors. IEEE J. Solid-State Circuits **24**(5), 1433–1439 (1989)
43. J.A. Croon, W. Sansen, H.E. Maes, *Matching Properties of Deep Sub-Micron MOS Transistors* (Springer, Berlin, 2005)

44. S. Decoutere, P. Wambacq, V. Subramanian, J. Borremans, A. Mercha, Technologies for (sub-) 45 nm analog/RF CMOS—circuit design opportunities and challenges, in *Proceedings of IEEE Custom Integrated Circuits, CICC* (2006), pp. 679–686
45. B. Razavi, *Design of Analog CMOS Integrated Circuits* (McGraw Hill, New York, 2001)
46. Y. Yausda, T.J. King Liu, C. Hu, Flicker-noise impact on scaling of mixed-signal CMOS with HfSiON. IEEE Trans. Electron Devices **55**(1), 417–422 (2008)
47. Z.M. Rittersma, M. Vertregt, W. Deweerd, S. van Elshocht, P. Srinivasan, E. Simoen, Characterization of mixed-signal properties of MOSFETs with high-k (SiON/HfSiON/TaN) gate stacks. IEEE Trans. Electron Devices **53**(5), 1216–1225 (2006)
48. T.L. Tewksbury, H.-S. Lee, Characterization, modeling, and minimization of transient threshold voltage shifts in MOSFETs. IEEE J. Solid-State Circuits **29**(3), 239–252 (1994)
49. A. Shanware, M.R. Visokay, J.J. Chambers, A.L.P. Rotondaro, J. McPherson, L. Colombo, Characterization and comparison of the charge trapping in HfSiON and HfO2 gate dielectrics, in *Technical Digest of International Electron Devices Meeting, IEDM* (2003), pp. 38.6.1–38.6.4
50. M. Fulde, D. Schmitt-Landsiedel, G. Knoblinger, Transient variations in emerging SOI technologies: modeling and impact on analog/mixed-signal circuits, in *Proceedings of IEEE International Symposium on Circuits and Systems, ISCAS* (2006)
51. P. Cuevas, A simple explanation for the apparent relaxation effect associated with hot-carrier phenomenon in MOSFETs. IEEE Electron Device Lett. **9**, 627–629 (1988)
52. L.J. McDaid, S. Hall, P.H. Mellor, W. Eccleston, Physical origin of negative differential resistance in SOI transistors. Electronics Lett. **25**, 827–828 (1989)
53. J. Jomaah, G. Ghibaudo, F. Balestra, J.L. Pelloie, Impact of self-heating effects on the design of SOI devices versus temperature, in *IEEE International SOI Conference Proceedings* (1995), pp. 114–115
54. N. Collaert, K. von Arnim, R. Rooyackers, T. Vandeweyer, A. Mercha, B. Parvais, L. Witters, A. Nackaerts, E. Altamirano Sanchez, M. Demand, A. Hikavyy, S. Demuynck, K. Devriendt, F. Bauer, I. Ferain, A. Veloso, K. De Meyer, S. Biesemans, M. Jurczak, Low-voltage 6T FinFET SRAM cell with high SNM using HfSiON/TiN gate stack, fin widths down to 10 nm and 30 nm gate length, in *Proc. of ICICDT* (2008)
55. A. Veloso, T. Hoffmann, A. Lauwers, H. Yu, S. Severi, E. Augendre, S. Kubicek, P. Verheyen, N. Collaert, P. Absil, M. Jurczak, S. Biesemans, Advanced CMOS device technologies for 45 nm node and below. Sci. Technol. Adv. Mater. **8**(3), 214–218 (2007)
56. G. Eneman, M. Jurczak, P. Verheyen, T. Hoffmann, A. De Keersgieter, K. De Meyer, Scalability of strained nitride capping layers for future CMOS generations, in *Proceedings of the 35th European Solid-State Device Research Conference, ESSDERC* (2005), pp. 449–452
57. R. Rooyackers, E. Augendre, B. Degroote, N. Collaert, A. Nackaerts, A. Dixit, T. Vandeweyer, B. Pawlak, M. Ercken, E. Kunnen, G. Dilliway, F. Leys, R. Loo, M. Jurczak, S. Biesemans, Doubling or quadrupling MuGFET fin integration scheme with higher pattern fidelity, lower CD variation and higher layout efficiency, in *Technical Digest of International Electron Devices Meeting, IEDM* (2006), pp. 1–4
58. C.H. Diaz, K. Goto, Y. Yasuda, C. Tsao, T. Chu, W. Lu, V. Chang, Y.T. Hou, Y.S. Chao, P.F. Hsu, C. Chen, K. Lin, J. Ng, W. Yang, C.H. Chen, Y.H. Peng, C.J. Chen, C.C. Chen, M. Yu, L.Y. Yeh, K.S. You, K.S. Chen, K.B. Thei, C.H. Lee, S.H. Yang, J.Y. Cheng, K.T. Huang, J.J. Liaw, Y. Ku, S.M. Jang, H. Chuang, M.S. Liang, 32 nm gate-first high-k/metal-gate technology for high performance low power applications, in *Technical Digest of IEEE International Electron Devices Meeting, IEDM* (2008), pp. 629–632
59. A. Dixit, A. Kottantharayil, N. Collaert, M. Goodwin, M. Jurczak, K. De Meyer, Analysis of parasitic S/D resistance in multiple-gate FETs. IEEE Trans. Electron Devices **52**(6), 1132–1140 (2005)
60. M. Guillorn, J. Chang, A. Bryant, N. Fuller, O. Dokumaci, X. Wang, J. Newbury, K. Babich, J. Ott, B. Haran, R. Yu, C. Lavoie, D. Klaus, Y. Zhang, E. Sikorski, W. Graham, B. To, M. Lofaro, J. Tornello, D. Koli, B. Yang, A. Pyzyna, D. Neumeyer, M. Khater, A. Yagishita, H. Kawasaki, W. Haensch, FinFET performance advantage at 22 nm: an AC perspective, in *Symposium on VLSI Technology, Digest of Technical Papers* (2008), pp. 12–13

61. K.R. Laker, W. Sansen, *Design of Analog Integrated Circuits and Systems* (McGraw Hill, New York, 1994)
62. P.E. Allen, D.R. Holberg, *CMOS Analog Circuit Design* (Oxford University Press, London, 2002)
63. V. Subramanian, B. Parvais, J. Borremans, A. Mercha, D. Linten, P. Wambacq, J. Loo, M. Dehan, N. Collaert, S. Kubicek, R.J.P. Lander, J.C. Hooker, S. Donnay, M. Jurczak, G. Groeseneken, W. Sansen, S. Decoutere, Device and circuit-level analog performance trade-offs: a comparative study of planar bulk FETs versus FinFETs, in *Technical Digest of International Electron Devices Meeting, IEDM* (2005), pp. 898–901
64. A. Chatterjee, K. Vasanth, D.T. Grider, M. Nandakumar, G. Pollak, R. Aggarwal, M. Rodder, H. Shichijo, Transistor design issues in integrating analog functions with high performance digital CMOS, in *Symposium on VLSI Technology, Digest of Technical Papers* (1999), pp. 147–148
65. B. Min, S.P. Devireddy, Z. Celik-Butler, W. Fang, A. Zlotnicka, H. Tseng, P. Tobin, Low frequency noise in submicrometer MOSFETs with HfO2, HfO2/Al2O3 and HfAlOx gate stacks. IEEE Trans. Electron Devices **51**(10), 1679–1687 (2004)
66. M. Fulde, A. Mercha, C. Gustin, B. Parvais, V. Subramanian, K. von Arnim, F. Bauer, K. Schruefer, G. Knoblinger, D. Schmitt-Landsiedel, Analog design challenges and trade-offs using emerging materials and devices, in *Proceedings of 33th European Solid-State Circuits Conference, ESSCIRC*, September 2007, pp. 123–126
67. X. Chen, S. Samavedam, V. Narayanan, K. Stein, C. Hobbs, C. Baiocco, W. Li, D. Jaeger, M. Zaleski, H. Yang, N. Kim, Y. Lee, D. Zhang, L. Kang, J. Chen, H. Zhuang, A. Sheik, J. Wallner, M. Aquilino, J. Han, Z. Jin, J. Li, G. Massey, S. Kalpat, R. Jha, N. Moumen, R. Mo, S. Kirshnan, X. Wang, M. Chudzik, M. Chowdhury, D. Nair, C. Reddy, Y.-W. Teh, C. Kothandaraman, D. Coolbaugh, S. Pandey, D. Tekleab, A. Thean, M. Sherony, C. Lage, J. Sudijohno, R. Lindsay, J.-H. Ku, M. Khare, A. Steegen, A cost effective 32 nm high-k/metal gate CMOS technology for low power applications with single-metal/gate-first process, in *Symposium on VLSI Technology, Digest of Technical Papers* (2008), pp. 88–89
68. M. Sato, Y. Sugita, T. Aoyama, Y. Nara, Y. Ohji, Impact of different nature of interface defect states on the NBTI and 1/f noise of high-k/metal-gate pMOSFETs between (100) and (110) crystal orientation, in *Symposium on VLSI Technology, Digest of Technical Papers* (2008), pp. 64–65
69. International Technology Roadmap for Semiconductors, http://www.itrs.net, 2007 edn. Chap. Radio Frequency and Analog/Mixed-Signal Technologies for Wireless Communications
70. T. Schulz, W. Xiong, C.R. Cleavelin, K. Schruefer, M. Gostkowski, K. Matthews, G. Gebara, R.J. Zaman, P. Patruno, A. Chaudhry, A. Woo, J.P. Colinge, Fin thickness asymmetry effects in multiple-gate SOI FETs (MuGFETs), in *Proceedings of the IEEE International SOI Conference* (2005), pp. 154–156
71. C. Gustin, A. Mercha, J. Loo, V. Subramanian, B. Parvais, M. Dehan, S. Decoutere, Stochastic matching properties of FinFETs. IEEE Electron Device Lett. **27**(10), 846–848 (2006)
72. M. Fulde, F. Kuttner, K. von Arnim, B. Parvais, A. Mercha, N. Collaert, R. Rooyackers, D. Schmitt-Landsiedel, G. Knoblinger, A 10-bit current-steering FinFET D/A converter, in *IEEE International SOI Conference Proceedings*, October 2008, pp. 95–96
73. H. Reisinger, O. Blank, W. Heinrigs, A. Muhlhoff, W. Gustin, C. Schlunder, Analysis of NBTI degradation- and recovery-behavior based on ultra fast VT-measurements, in *Reliability Physics Symposium Proceedings, 2006. 44th Annual., IEEE International* (2006), pp. 448–453
74. S. Pae, M. Agostinelli, M. Brazier, R. Chau, G. Dewey, T. Ghani, M. Hattendorf, J. Hicks, J. Kavalieros, K. Kuhn, M. Kuhn, J. Maiz, M. Metz, K. Mistry, C. Prasad, S. Ramey, A. Roskowski, J. Sandford, C. Thomas, J. Thomas, C. Wiegand, J. Wiedemer, BTI reliability of 45 nm high-K + metal-gate process technology, in *Proceedings of 46th Annual International Reliability Physics Symposium, Phoenix* (2008), pp. 352–357
75. B.M. Tenbroek, M.S.L. Lee, W. Redman-White, R.J.T. Bunyan, M.J. Uren, Impact of self-heating and thermal coupling on analog circuits in SOI CMOS. IEEE J. Solid-State Circuits **33**(7), 1037–1046 (1998)

76. W. Molzer et al., Self heating simulation of multi-gate FETs, in *Proceedings of 36th European Solid-State Device Research Conference, ESSDERC* (2006), pp. 311–314
77. B.M. Tenbroek, M.S.L. Lee, W. Redman-White, R.J.T. Bunyan, M.J. Uren, Self-heating effects in SOI MOSFETs and their measurement by small signal conductance techniques. IEEE Trans. Electron Devices **43**(12), 2240–2249 (1996)
78. L. Bertolissi, Modeling and simulation of self-heating effects in deep sub-micron silicon on insulator (SOI) technologies for analog circuits. M.Sc. Thesis, Universita Degli Studi di Udine (2004)
79. B. Razavi, *RF Microelectronics* (Prentice Hall, New York, 1998)
80. T.L. Lee, *The Design of CMOS Radio-Frequency Integrated Circuits* (Cambridge University Press, Cambridge, 2004)
81. A. Hajimiri, T.H. Lee, A general theory of phase noise in electrical oscillators. IEEE J. Solid State Circuits **33**(2), 179–194 (1998)
82. K.C. Hsieh, P. Gray, D. Senderowicz, D.G. Messerschmitt, A low-noise chopper-stabilized switched capacitor filtering technique. IEEE J. Solid State Circuits **16**(6), 708–715 (1981)
83. R. Gregorian, G.C. Temes, *Analog MOS Integrated Circuits for Signal Processing* (Wiley, New York, 1986)
84. I. Bloom, Y. Nemirovsky, 1/f noise reduction of metal-oxide-semiconductor transistors by cycling from inversion to accumulation. Appl. Phys. Lett. **58**(15), 1664–1666 (1991)
85. E.A.M. Klumperink, S.L.J. Gierkink, A.P. van der Wel, B. Nauta, Reducing MOSFET 1/f noise and power consumption by switched biasing. IEEE J. Solid State Circuits **35**(7), 994–1001 (2000)
86. J. Koh, R. Thewes, D. Schmitt-Landsiedel, R. Brederlow, A circuit design-based approach for 1/f-noise reduction in linear analog CMOS ICs, in *Symposium on VLSI Technology, Digest of Technical Papers* (2004), pp. 222–225
87. J. Koh, D. Schmitt-Landsiedel, R. Thewes, R. Brederlow, A complementary switched MOSFET architecture for the 1/f noise reduction in linear analog CMOS ICs. IEEE J. Solid State Circuits **42**(6), 1352–1361 (2007)
88. D. Siprak, N. Zanolla, M. Tiebout, P. Baumgartner, C. Fiegna, Reduction of low-frequency noise in MOSFETs under switched gate and substrate bias, in *Proceedings of the 38th European Solid-State Device Research Conference, ESSDERC* (2008), pp. 266–269
89. D. Siprak, M. Tiebout, P. Baumgartner, Reduction of VCO phase noise through forward substrate biasing of switched MOSFETs, in *Proceedings of 34th European Solid State Conference, ESSCIRC* (2008), pp. 326–329
90. M. Ertuerk, T. Xia, W. Clark, Gate voltage dependence of MOSFET 1/f noise statistics. IEEE Electron Device Lett. **28**(9), 812–814 (2007)
91. B. Razavi, *Principles of Data Conversion System Design* (IEEE Press, New York, 1995)
92. F. Kuttner, A 1.2 V 10b 20MSample/s non-binary successive approximation ADC in 0.13 um CMOS, in *IEEE International Solid-State Circuits Conference, Digest of Technical Papers*, vol. 1 (2002), pp. 176–177
93. R. Schreier, G.C. Temes, *Understanding Delta-Sigma Data Converters* (Wiley, New York, 2004)
94. S.R. Norsworthy, R. Schreier, G.C. Temes, *Delta-SIGMA Data Converters: Theory, Design, and Simulation* (IEEE Press, New York, 1996)
95. G.D.J. Smit, A.J. Scholten, N. Serra, R.M.T. Pijper, R. van Langevelde, A. Mercha, G. Gildenblat, D.B.M. Klaassen, PSP-based compact FinFET model describing DC and RF measurements, in *Technical Digest of International Electron Devices Meeting, IEDM*, December 2006, pp. 1–4
96. W. Sansen, *Analog Design Essentials* (Springer, Berlin, 2006)
97. K.E. Kuijk, A precision reference voltage source. IEEE J. Solid State Circuits **SC-8**, 222–226 (1973)
98. S.M. Sze, *Physics of Semiconductor Devices* (Wiley, New York, 1981)
99. U. Feldmann, Diode model in TITAN V006. Technical Report, Infineon Technologies, TITAN group (2005)

100. M. Wirnshofer, Evaluation of voltage reference circuits using novel devices. M.S. Thesis, Technische Universität München (2007)
101. H. Banba, H. Shiga, A. Umezawa, T. Miyaba, T. Tanzawa, S. Atsumi, K. Sakui, A CMOS bandgap reference circuit with sub-1-V operation. IEEE J. Solid State Circuits **34**(5), 670–674 (1999)
102. M. Fulde, M. Wirnshofer, G. Knoblinger, D. Schmitt-Landsiedel, Design of low-voltage bandgap reference circuits in multi-gate CMOS technologies, in *Proceedings of IEEE International Symposium on Circuits and Systems, ISCAS* (2009), pp. 234–237
103. J.A. Schoeff, An inherently monotonic 12 bit DAC. IEEE J. Solid-State Circuits **14**, 904–911 (1979)
104. A. van den Bosch, M.A.F. Borremans, M.S.J. Steyaert, W. Sansen, A 10-bit 1-GSample/s Nyquist current-steering CMOS D/A converter. IEEE J. Solid-State Circuits **36**(3), 315–324 (2001)
105. A. van den Bosch, M. Steyaert, W. Sansen, SFDR-bandwidth limitations for high speed high resolution current steering CMOS D/A converters, in *Proceedings of the IEEE International Conference on Electronics, Circuits and Systems, ICECS*, vol. 3 (1999), pp. 1193–1196
106. A. Van den Bosch, M. Steyaert, W. Sansen, An accurate statistical yield model for CMOS current-steering D/A converters, in *Proceedings of IEEE International Symposium on Circuits and Systems, ISCAS*, vol. 4 (2000), pp. 105–108
107. W.-S. Chou, S.-C. Yang, F.-L. Hsueh, H.-C. Huang, C.-J. Hsiao, A low-cost triple-channel 10-bit 250 MHz DAC IP in 65 nm CMOS process, in *Proceedings of IEEE International Symposium on Circuits and Systems, ISCAS* (2007), pp. 3594–3597
108. D.C. Lee, Analysis of jitter in phase-locked loops. IEEE Trans. Circuits Syst. II Analog Digit. Signal Process. **49**(11), 704–711 (2002)
109. B. Razavi, *Monolitic Phase-Locked Loops and Clock Recovery Circuits* (IEEE Press, New York, 1996)
110. S. Williams, H. Thompson, M. Hufford, E. Naviasik, An improved CMOS ring oscillator PLL with less than 4 ps RMS accumulated jitter, in *Proceedings of IEEE Custom Integrated Circuits Conference* (2004), pp. 151–154
111. H. Arora, N. Klemmer, J.C. Morizio, P.D. Wolf, Enhanced phase noise modeling of fractional-N frequency synthesizers. IEEE Trans. Circuits Syst. I Regul. Pap. **52**(5), 379–395 (2005)
112. Z. Cao, Y. Li, S. Yan, A 0.4 ps-RMS-jitter 1–3 GHz ring-oscillator PLL using phase-noise preamplification. IEEE J. Solid State Circuits **43**(9), 2079–2089 (2008)
113. N. DaDalt, C. Sandner, Private communication
114. R. Nonis, Phase noise modelling in phase lock loop frequency synthesizers. M.S. Thesis, University of Udine (2002)
115. L. Bizjak, N. DaDalt, P. Thurner, R. Nonis, P. Palestri, L. Selmi, Comprehensive behavioral modeling of conventional and dual-tuning PLLs. IEEE Trans. Circuits Syst. I Regul. Pap. **55**, 1628–1638 (2008)
116. A. Hajimiri, S. Limotyrakis, T.H. Lee, Jitter and phase noise in ring oscillators. IEEE J. Solid State Circuits **34**(6), 790–804 (1999)
117. G. Knoblinger, M. Fulde, D. Siprak, U. Hodel, K. von Arnim, T. Schulz, C. Pacha, U. Baumann, W. Marshall, A. Xiong, C.R. Cleavelin, P. Patruno, K. Schruefer, Evaluation of FinFET RF building blocks, in *IEEE International SOI Conference Proceedings*, October 2007, pp. 39–40
118. M. Fulde, K. von Arnim, C. Pacha, F. Bauer, C. Russ, D. Siprak, W. Xiong, A. Marshall, C.R. Cleavelin, K. Schruefer, D. Schmitt-Landsiedel, G. Knoblinger, Advances in multi-gate MOSFET circuit design, in *Proceedings of the 14th IEEE International Conference on Electronics, Circuits and Systems, ICECS*, December 2007, pp. 186–189
119. M. Tiebout, *Low Power VCO Design in CMOS* (Springer, Berlin, 2005)
120. M. Dehan, B. Parvais, V. Subramanian, C. Gustin, J. Loo, A. Mercha, S. Decoutere, Characterization, modeling, and optimization of FinFET MOS varactors, in *Topical Meeting on Silicon Monolithic Integrated Circuits in RF Systems* (2007), pp. 28–31

121. J. Janssens, M. Steyaert, H. Miyakawa, A 2.7 V CMOS broadband low noise amplifier, in *Symposium on VLSI Circuits, Digest of Technical Papers* (1997), pp. 87–88
122. M. Tiebout, E. Paparisto, LNA design for a fully integrated CMOS single chip UMTS transceiver, in *Proceedings of the 28nd European Solid-State Circuits Conference, ESSCIRC* (2002), pp. 835–838
123. B. Parvais, C. Gustin, V. de Heyn, J. Loo, M. Dehan, V. Subramanian, A. Mercha, N. Collaert, R. Rooyackers, M. Jurczak, P. Wambacq, S. Decoutere, Suitability of FinFET technology for low-power mixed-signal applications, in *IEEE International Conference on Integrated Circuit Design and Technology, ICICDT'06* (2006), pp. 1–4
124. P. Wambacq, B. Verbruggen, K. Scheir, J. Borremans, V. De Heyn, G. Van der Plas, K. Mercha, Bb. Parvais, V. Subramanian, M. Jurczak, S. Decoutere, S. Donnay, Analog and RF circuits in 45 nm CMOS and below: planar bulk versus FinFET, in *Proceedings of 33rd European Solid State Conference, ESSCIRC* (2006), pp. 54–57
125. B.M. Tenbroek, W. Redman-White, M.S.L. Lee, R.J.T. Bunyan, M.J. Uren, K.M. Brunson, Characterization of layout dependent thermal coupling in SOI CMOS current mirrors. IEEE Trans. Electron Devices **43**(12), 2227–2232 (1996)
126. I.M. Filanovsky, A. Allam, Mutual compensation of mobility and threshold voltage temperature effects with applications in CMOS circuits. IEEE Trans. Circuits Syst. I Fundam. Theory Appl. **48**, 876–884 (2001)
127. M. Fulde, J.P. Engelstaedter, G. Knoblinger, D. Schmitt-Landsiedel, Analog circuits using FinFETs: benefits in speed-accuracy-power trade-off and simulation of parasitic effects. Adv. Radio Sci. Kleinheubacher Berichte 2006 **5**, 285–290 (2007)
128. V. Subramanian, A. Mercha, B. Parvais, J. Loo, C. Gustin, N. Collaert, M. Jurczak, G. Groeseneken, W. Sansen, S. Decoutere, Optimization of FinFET geometries for analog performance, in *Proceedings of ULIS* (2006)
129. C. Kienmayer, An integrated 17 GHz receiver in 0.13 um CMOS for wireless applications. Ph.D. Thesis, Technische Universität Wien (2004)
130. A. Bargagli-Stoffi, *Ultra Low-Voltage, Low-Power Amplifiers in Deep-Submicrometer CMOS* (Shaker, Maastricht, 2006)
131. T. Nirschl, *Circuit Applications of the Tunneling Field Effect Transistor (TFET)* (Shaker, Maastricht, 2008)
132. M. Fulde, A. Heigl, M. Weis, M. Wirnshofer, K.V. Arnim, T. Nirschl, M. Sterkel, G. Knoblinger, W. Hansch, G. Wachutka, D. Schmitt-Landsiedel, Fabrication, optimization and application of complementary multiple-gate tunneling FETs, in *Proceedings of 2nd IEEE International Nanoelectronics Conference*, March 2008, pp. 946–951
133. Q. Zhang, W. Zhao, A. Seabaugh, Low-subthreshold-swing tunnel transistors. IEEE Electron Device Lett. **27**(4), 297–300 (2006)
134. K. Boucart, A. Iounescu, Threshold voltage in tunnel FETs: physical definition, extraction, scaling and impact on IC design, in *Proceedings of the 37th European Solid-State Device Research Conference, ESSDERC* (2007), pp. 299–302
135. G.A.M. Hurkx, D.B.M. Klaassen, M.P.G. Knuvers, A new recombination model for device simulation including tunneling. IEEE Trans. Electron Devices **39**, 331–338 (1992)
136. T. Nirschl, P.-F. Wang, W. Hansch, D. Schmitt-Landsiedel, The tunneling field effect transistors (TFET): the temperature dependence, the simulation model, and its application, in *Proceedings of the International Symposium on Circuits and Systems, ISCAS*, vol. 3 (2004), pp. 713–716
137. M. Fulde, A. Wachutka, G. Heigl, D. Schmitt-Landsiedel, Complementary multi-gate tunneling FETs: fabrication, optimization and application aspects. Int. J. Nanotechnology **6**(7/8), 628–639 (2009)
138. A. Heigl, G. Wachutka, Simulation of advanced tunneling devices, in *International Conference on Advanced Semiconductor Devices and Microsystems, ASDAM* (2006), pp. 121–124

139. A. Heigl, G. Wachutka, Optimization of vertical tunneling field-effect transistors, in *Proc. of ULIS* (2007), pp. 133–136
140. W.Y. Choi, J.Y. Song, J.D. Lee, Y.J. Park, B.-G. Park, 70-nm impact-ionization metal-oxide-semiconductor (I MOS) devices integrated with tunneling field-effect transistors (TFETs), in *Technical Digest of International Electron Devices Meeting, IEDM* (2005), pp. 955–958